Harold K. Steen

Forest Service Research

Finding Answers to Conservation's Questions

Forest History Society   Durham, North Carolina   1998

© 1998 by the Forest History Society
Printed in the United States of America
All rights reserved.

The Forest History Society is a nonprofit,
educational institution dedicated to the
advancement of historical understanding
of man's interaction with the forest environment. It was established in 1946.
Interpretations and conclusions in FHS
publications are those of the authors; the
institution takes responsibility for the
selection of topics, the competence of the
authors, and their freedom of inquiry.

This publication was developed in
cooperation with the USDA Forest Service,
Southeastern Forest Experiment Station,
Asheville, North Carolina, and the Forest
Service History Program.

Cover and title page illustration: The
blinkometer, used in forest fire research,
measures the moisture content of logs at the
Priest River Experiment Station in Idaho.
(U.S. Forest Service photo.)

Book design by Teresa Smith Perrien

Library of Congress
Cataloguing-in-Publication Data

Steen, Harold K.
    Forest Service research : finding answers
to conservation's questions / by Harold K.
Steen.
        p.   cm.
    "In cooperation with USDA – Forest
Service."
    Includes bibliographic references (p.   )
and index.
    ISBN 0-89030-056-9
    1. United States. Forest Service Research.
2. Forests and forestry—Research—United
States. I. Forest History Society. II. United
States. Forest Service. III. Title.
IN PROCESS
634.9'072073—dc21   98-8454
                                                    CIP

CONTENTS

Preface, v

PART *1* THE EARLY YEARS, 1

Agriculture Takes The Lead in Science, 2
Moving Beyond Compilation, 4
Research Under Pinchot, 5
Forest Products Laboratory, 8
Central Investigative Committee, 8
Forest/Flood Study, 9

PART *2* RESEARCH STRIVES FOR INDEPENDENCE, 10

Earle H. Clapp—Research Architect, 11
World War I, 12
Pulling the Research Program Together, 14
Range Research, 17
Research Natural Areas, 18
Research Councils, 19
McSweeney-McNary Act 19,
A New Deal for Research, 22
Copeland Report, 22
Forestland Taxes, 24
Douglas-Fir Study, 25
Range Controversy, 26
Statistical Training, 27
Other Depression Research, 29
Clapp Era Ends, 32
Clarence Forsling and World War II, 33
International Forestry, 34
The Kotok Years and Postwar Research, 35

PART *3* RESEARCH EXPANSION BEGINS, 39

Verne L. Harper Meets the Congress, 39
Relations with the USDA, 40
Cultivating Congress, 42
A Variety of Programs, 43

Timber Supply, 47
Forest Research Advisory Committee, 48
Fire and Water Research, 50
Forest Products Laboratory, 51
Administrative Advances, 52
Forestry Research Plan, 53
McIntire-Stennis Law, 55

PART 4 RESEARCH AND THE ENVIRONMENT, 57
Silent Spring in a Noisy Decade, 57
Minorities in the Workplace, 57
Research at Mid Decade, 58
Administrative Issues, 61
The Environmental Decade, 62
At the Cutting Edge, 63
A Quality Environment, 64
Changing Times, 65
Dickerman Cools Things Off, 68
The Buckman Administration, 71
A New Research Act, 73
Change of Administration, 73
Biotechnology, 75
The Illusive Ivory Tower, 76
International Program, 76
Research Independence, 77
More with Less, 78
Ecosystem Research, 79
Strategy for the 90's, 79
A Sampling of Research, 81
The Owl and the Woodpecker, 81
Mandate for Change, 82
The Voyage Begins, 83

Selected References, 85
Index, 89

PREFACE

The participating agreement between the USDA-Forest Service and the Forest History Society refers to a "one-hundred page book" on the history of the agency's research efforts for more than a century. Although no one has assumed that the length needed to be precisely as stated in the agreement, it did mean that the book was to be a synthesis, an overview—much more has been left out than included. The goal has been to examine key events in relative detail but provide only samples of daily activities. The full story is robust; it is hoped that this glimpse captures it adequately.

To historians the context—the whys—is an essential ingredient for any story. Thus what follows is much more than a listing of events. We also see the people—frail humans all—who made it happen, as well as the broader themes that influenced both priorities and outcomes.

Many have contributed to this effort; especially the agency itself as the experiment stations and many experimental forests have compiled their individual histories. Many stations also responded to requests to nominate their "top 10" scientists. Too, the Forest Service over the years has sponsored in depth interviews with its key players, including those in research. Not a few retirees have provided both published and unpublished renditions of their careers. Added to that are the research publications that have appeared in legion. Finally, the author was a junior scientist for four years at the Pacific Northwest Forest and Range Experiment Station during the early 1960s and to that extent has a sense of research "culture."

Former deputy chiefs V. L. Harper, M. B. Dickerman, Robert E. Buckman, John H. Ohman, and Jerry A. Sesco, former station director Charles W. Philpot, and Judy Cook, secretary to Buckman, Ohman, and Sesco, were kind enough to read an early draft and supply the author with critiques. Their suggestions, general observations, and at times corrections have been very useful.

"Recently, the Forest Service has embarked on what could be called a voyage 'beyond the maps.' Ecosystem management, sustainability, biodiversity, forest health—these concepts are taking the agency outside its traditional boundaries. . . ." Thus begins the report from the 1995 Forest Service Science and Policy Roundtable, where "integrating science into land management decision making was the key topic." Clearly, by the century's end, the Forest Service research program was seen as an equal partner to the agency's management and outreach activities, a condition that did not hold as the century began. But the story begins much earlier.

PART THE EARLY YEARS

Cumbersome, official language aside, the United States was still a fairly young nation when forestry research became visible. "The President is desirous of causing to be introduced into the United States all such trees and plants from other countries, not heretofore known in the United States. . ." Thus begins an 1827 Treasury Department circular to all U.S. consuls to seek out species that showed promise for transplanting; "forest trees useful for timber" were on the list of things to look for. The Navy would provide free shipment, and specialists would undertake trial cultivation. One such specialist was a Dr. Henry Perrine, who began a promising hemp plantation in Florida. However, the experiment was never completed, as Perrine was killed during the Seminole war of 1840 and his records destroyed.

A year after President John Quincy Adams' 1827 request, another Florida experiment took place, this one to attempt cultivation of liveoak to assure future supplies of ship timbers needed for the Navy. Adams himself had germinated acorns in tubs on the White House back porch and believed it could

work in the field, asking the Navy to see what could be done. Shortly thereafter, Andrew Jackson defeated Adams' bid for reelection and quickly can-
celed the liveoak experiment, as well as other programs. Adams' great intel-
lect and erudition notwithstanding, he was having difficulty getting American forest research up and running.

Solid science was minimal. Most of what was known about American tree
species stemmed from plant explorers—often European—who classified as
they trekked the mountains and paddled the rivers. The most-used reference
was *Sylva: A Discourse on Trees*, authored by John Evelyn and first published
by the Royal Society of London in 1664. Adams' diary reveals that the president, too, had a well-used copy of *Sylva*, and it was his authority on trees.

An examination of Evelyn's classic quickly shows he largely based the
work on the much earlier writings of Pliny, Theophrastis, Cato, and other
Greek and Roman ancients. Thus, we see American forest science during the
early part of the nineteenth century being strongly influenced by those who
preceded by as much as two millennia. For a nation so mightily endowed with
forests, it was time for American scientists to take a fresh look.

The time would come after the century passed its midpoint; in 1864
George Perkins Marsh published his still-influential *Man and Nature: The
Earth as Modified by Human Action*. Marsh had scoured the natural literature
in many languages to synthesize a masterpiece that clearly showed the cumu-
lative effect of human misuse of forests, forage, soil, and water. His scene was
largely European, but the lesson applied to the still-evolving America—use
but do not abuse. Promptly and from then on, Marsh was cited in government
reports, on the floor of Congress, and by scientists who could understand what
the problem was: America was expanding rapidly without apparent concern
about the impact on land and its resources.

Agriculture Takes the Lead in Science

The very next year, the annual report of the commissioner of agriculture included a report by the Reverend Frederick Starr, who implored, "let exten-
sive, protracted, and scientific experiments in the propagation and cultivation
of forest trees be established." More and more often, scientists would be asked
to turn their attention to forested lands. One of the more significant responses
would come from a New York State physician, Franklin B. Hough.

In 1873 Hough made a presentation on the forestry responsibilities of government to the American Association for the Advancement of Science. The
august body of scientists responded by petitioning Congress on the urgent

need to gather and distribute forestry facts. Congress was not all that impressed, but three years later it appropriated $2,000 to gather forestry information. However, the appropriation was for Agriculture—not the Department of the Interior as proposed. The commissioner of agriculture asked Hough to do the job; the appropriation was subsequently renewed and increased, and Hough would be involved for nearly a decade.

In addition to this accumulation of knowledge, Hough began the tradition of forestry as part of agricultural activity where scientific investigations were already established.

Hough compiled three volumes and contributed to a fourth entitled *A Report Upon Forestry*. Volume I (1877) contained information on the general distribution of U.S. forests, benefits of sowing seeds versus planting seedlings, cottonwood growth rates, soil properties, naval stores, insects, disease, and shelterbelts. Also included was a section on John Evelyn's [1662] advice on tree planting, showing how this much earlier work was still seen as a valid authority. Put differently, *Sylva* had yet to be replaced.

Hough's second volume (1880) looked almost entirely at state forestry activities and import/export statistics, and volume IV (1884), compiled by Nathanial Egleston who was Hough's lusterless successor, offered status reports on various regions. However, volume III (1882) zeroed in on forest science with a twenty-page section, "Experimental stations for forest culture." Hough began, "There is no kind of cultivation that involves a wider range of capabilities than that of forest trees, . . ." He pointed out that information gathered in one area did not necessary apply to another; regional experiment stations were needed to deal with ecological and species diversity. He proposed stations in midwestern prairies, at the edge of arid regions, in southern California, and in Florida. He further proposed that another Florida experiment station was needed to focus on naval stores, products extracted from pine resin.

Others were working on forest science, too. At Harvard's Arnold Arboretum, Charles S. Sargent was compiling the fourteen-volume *Silva of North America* that would be published serially between 1890 and 1902. Earlier, Sargent had prepared a volume for the 1880 census; his *Report on the Forests of North America* appeared in 1884. In it, he reported on dendrology and tree species distribution and wood properties. The arboretum itself contributed taxonomic and silvical knowledge.

A less obvious but still profoundly important element of forest science was underway at the U.S. Geological Survey. As part of its world-class studies of landform creation, the agency studied the role of forests on waterflow. This was an especially important topic, as Congress was seeking rationale for re-serving the forested watersheds of the West. Two of the key scientists were Arnold Hague and W. J. McGee, whose names are writ large in conservation history.

Moving Beyond Compilation

An even larger name was that of Bernhard Eduard Fernow, a German for-ester who had immigrated to the United States in 1876. He quickly assumed a leadership role in the newly founded American Forestry Association and in 1886 was appointed chief of the Division of Forestry in the Department of Agriculture. He would hold that position for twelve years, longer than any subsequent chief. He was active in the American Association for the Ad-vancement of Science, he testified to Congress on matters of forest science, and he wrote both scientific and popular articles plus two books that are still use-ful.

Fernow left federal service in 1898 to become dean of the newly founded forestry school at Cornell University. At the request of the secretary of agri-culture, Fernow had assembled a 450-page *Report upon the Forestry Investiga-tions of the US Department of Agriculture, 1877–1898*, published by the depart-ment in 1899. Included in the document is a financial statement; the total Division of Forestry budget for the dozen years was $247,000, of which $174,000 had been allocated to investigations, leaving $73,000 for administra-tive costs. When compared to the division publication record, Fernow noted, they had disbursed information for $24 per published page. He thought that to be a bargain.

Fernow listed twenty-three major publications, twenty informational circulars, and a variety of reports. Some of the works were lengthy; one on the use of metal for railroad ties was 363 pages. E. E. Russell Tratman was author of this major work and others on means to conserve wood by using other materials. Charles Mohr wrote on southern pine, George B. Sudworth on dendrology, Frederick V. Coville on sheep grazing, and Filibert Roth on Wis-consin forestry conditions.

Sudworth was one of the first associates that Fernow had recruited in 1876. Trained in botany, Sudworth contributed some of the most enduring work of the division. His 1897 *Nomenclature of the Arborescent Flora of the United States* analyzed six thousand names applied to five hundred forest species. His later checklists of trees remained in print for a half century. He remained active with the agency through the 1920s and among other tasks taught tree identification to the women of the Washington Office. His indefatigable and meticulous research established a dendrological tradition that Elbert Little, also of long tenure, would carry on beyond his retirement in the 1970s, when the Linnaean system of classification was replaced by molecular analysis.

Another of Fernow's frequent collaborators, Filibert Roth, assisted with technology transfer from Europe. Looking back, we can see that their joint work on wood decay—published in Fernow's annual reports of the Division of Forestry—represent the origins of American forest pathology.

Fernow's interests were broad—forest biology, timber physics, and soil— and his reports and other publications addressed aspects of the whole field of forest science. Timber physics studies, at times in collaboration with a Washington University testing laboratory in St. Louis, would lead not only to more efficient wood use but provide tips to the forest manager on which species to grow and optimum size tree. He persuaded lumber companies to donate beams to be tested and railroads to contribute transportation charges, thus extending his meager research budget.

Forest influences, a field that retained its identity until the middle of the twentieth century, was another of Fernow's interests. In collaboration with others, he published in 1893 a 197-page study of forest/water relationships, and he often drew upon this work in subsequent reports that influenced Congress as it debated the purposes of the national forests. A related assignment was rain-making, a study that he felt was a bit silly and managed to shunt off to the Army Signal Corps. He ended the fiscal year with his rain-making allocation unspent.

The broader context of forest research remained important. In 1868 the Division of Botany began in Agriculture (received cabinet rank in 1889), which studied range grasses among its larger mission. In 1888 Congress had passed the Hatch Act, providing federal funds for establishing state agricultural experiment stations. Many stations were adjunct to the state university, where there might be an arboretum, a program to distribute tree seedlings, or work on breeding. In 1895 the new Division of Agrostology provided additional emphasis on range research. Controversy in Oregon over grazing led to Coville's landmark study that permitted grazing but under regulation. Momentum and infrastructure for forest research were building.

Research Under Pinchot

Most Division of Forestry changes in 1898 were substantial, but some were more a matter of appearance. In that year, Gifford Pinchot succeeded Fernow as chief of the division. Pinchot's dynamic leadership, his political finesse, and his bureaucratic skills in changing the small, behind-the-scenes division into an institution that would become the centerpiece of the unfolding conservation movement is a much-told story. What is less known is the important role of research in the evolving agency.

Pinchot's vision of the agency and how to achieve it was different than Fernow's. Pinchot saw need and opportunity to manage the growing [since 1891] national forest [forest reserve] system, which he would realize in 1905 with the transfer of the forests to the jurisdiction of the Department of Agriculture. To counter strong western opposition to federal land management, Pinchot constantly emphasized the practicality of his approach and that of his agency. After 1905 western political cartoonists would refer to him as "Professor" Pinchot, an epithet that he vigorously tried to deflect by demonstrating he was not a theorist. An important part of that demonstration had been to shift rhetoric; from 1898 onward, "research" was no longer on the agency's agenda, but "investigations" were. This minor deception would remain in place until 1915. Lack of precise definition makes quantification risky, but perhaps as much as 25 percent of Pinchot's budget was research related.

In his 1901 report, Pinchot listed the achievements of the Section of Special Investigations: "studies of commercial trees, forest fires, grazing, log scales, forests and water supply, compilation of forest histories, and the investigation of forest products." Related to investigations was the agency library that housed 110 "bound volumes" and 1,300 pamphlets. The photo collection numbered 4,968 prints. Subsequent reports included ever-larger sections on "investigations," and the numbers of books and photos also increased.

The people are impressive, too. In 1904 we see that George B. Sudworth is chief of dendrology and Herman von Schrenk is chief of forest products. Von Schrenk, associated with the Missouri Botanical Gardens, reflects Pinchot's broad network of cooperators. Von Schrenk was also involved with the 1904 St. Louis Exposition, where the Forest Service [Bureau of Forestry] had set up a small timber-preserving plant. Scientists made six thousand tests of treated and natural woods with results published in a circular entitled, "Experiments on the Strength of Treated Timber."

Other scientific collaborators were Professor W. L. Jepson in Berkeley who was studying tanoak, H. D. Tiemann of the Yale Forest School who was measuring the effect of moisture content on the strength of southern pine, Samuel J. Record was in Montana working on ponderosa pine, Raphael Zon was involved with a commercial tree study in Tennessee, H. S. Betts was in Washington, D.C. testing southern pine timbers, and Charles Herty had begun his four-decade work on products derived from southern pine. In all, forty-two individual investigations were listed.

As the agency, now the USDA-Forest Service, evolved administratively following the 1905 forest reserve transfer from the Department of the Interior, research also evolved and became more identifiable. As part of his effort to decentralize the agency, Pinchot in 1908 divided the national forests (all in the West) into six inspection districts (since 1930 called regions). The same year saw establishment of the first experiment station in Fort Valley, Arizona, with Gus Pearson as director.

The plan was to establish stations in each silvicultural region, including those in the East. Years later, Pearson looked back at the founding of the Fort Valley Station and remembered riding by horse and buggy from Flagstaff. He accompanied Raphael Zon from the Washington Office and the supervisor of the Coconino National Forest. A thunderstorm caused normally dry Rio de Flag to flood, and it was "running a hundred yards wide with a fluid whose color and consistency told plainly that the country was going to the dogs even in that early day." When they arrived at the selected site a half-mile beyond the flood, Zon said as he often did for dramatic effect, "Here we shall plant the tree of research."

The pattern was soon thereafter repeated at other sites, with Zon personally accompanying the local Forest Service officers to approve the location. In the case of Fort Valley, Zon's aide, Samuel T. Dana, had selected the site. Each station had to be effective administratively, such as convenient access to trans-

Gus Pearson photographed these administrative quarters at Fort Valley Experiment Station in 1912, when he served as the station's first director. (U.S. Forest Service photo.)

7

portation, and needed to be adjacent to forest types that were typical of the area.

"Investigations" was divided and renamed in arcane fashion; after 1907 we see Raphael Zon as chief of the Office of Silvics and William L. Hall heading the Branch of Products. Samuel T. Dana was assisting Zon, and Carlos Bates was beginning his very long term study of shelterbelts. Products studies included naval stores and related wood chemistry and testing beams sawn from a variety of species. A decade earlier, the secretary of agriculture had suspended Fernow's timber-testing studies as "inappropriate," but Pinchot not only reinstated them, he greatly expanded the efforts.

Forest Products Laboratory

The agency took advantage of timber-testing laboratories—some developed with Forest Service cooperation—at Yale, University of California, Purdue University, the E. P. Burton Lumber Company in South Carolina, and many other universities. Betts and McGarvey Cline used a variety of laboratories, but while they were working at Burton Lumber in 1906, Cline came up with the idea for a central facility. Pinchot would agree, and in 1910 the Forest Products Laboratory opened in Madison, Wisconsin, in cooperation with the state university. The Madison lab quickly became a world-class operation, contributing not only to basic understanding of wood chemistry and structure but also many commercially successful processes. It was also central to the conservation movement; better utilization meant less waste, and fewer trees would contribute more products.

Central Investigative Committee

President Taft fired the insubordinate Pinchot before the Forest Products Laboratory opened its doors, but the new chief, Henry Graves, continued most policies and priorities. In 1911 he issued a new manual that instructed agency personnel on the proper means for research: "Forest investigation of a thorough and systematic character and conducted with scientific accuracy form an important part of the work of the Forest Service." All members of the agency were directed to cooperate.

Graves wanted more structure, and in 1912 he established a Central Investigative Committee consisting of Raphael Zon, range specialist J. T. Jardine, and Carlisle P. Winslow representing products. The central committee worked with branch chiefs to set priorities and provided overall supervision of the research program. Each of the six districts were to set up similar committees, which would assist the district forester. Also, experiment station heads were to report to the district forester. Projects needed to be carefully designed and approved regionally.

All was not yet decentralized. Book purchases at either the district or national forest level had to be approved in Washington, D.C. Rangers were encouraged to borrow such books from their forest supervisor for reading "at home."

One ongoing research goal is the dissemination of findings, and in 1913 Graves reported to the secretary that he was issuing the *Review of Forest Service Investigations* from time to time to keep "the entire technical work force informed." The reports were "designed to improve investigative methods, avoid duplication, and stimulate interest in research work." The two-part report included about one hundred fifty pages of information on dendrology, grazing, silviculture, and products with details of many studies within each basic category. Of course, much of the work would eventually be published in full, but the purpose here was to provide a status report aimed at a broad readership.

The research infrastructure had grown substantially in the five short years since the first experiment station had opened in Fort Valley. By 1913 there was of course the Forest Products Laboratory but also experiment stations in Idaho, Minnesota, Colorado, Washington, and Utah, plus the dendrological and seed testing laboratories in Washington, D.C., and two regional products labs in Wisconsin and Washington state.

Forest/Flood Study

One early station, the Wagon Wheel Gap Experiment Station on Colorado's Rio Grande National Forest, was born in controversy. One of the primary purposes of the national forests was to "maintain favorable conditions of water flow," and the only serious attempt to measure the influence of forests on streamflow was the Emmenthal study in Switzerland. Thus, two drainages in southern Colorado were selected for long term study.

Wagon Wheel Gap received instant prominence that would last a decade. During the long debate that led to the 1911 passage of the Weeks Act, authorizing federal purchase of certain forested watersheds for addition to the national forest system, the Forest Service had repeatedly gone on record that forests were essential regulators of floods. The Army Corps of Engineers vigorously countered that levies and dams were the only way to control water.

Wagon Wheel Gap was established to validate the testimony of Pinchot and others. Unfortunately, the results were less conclusive than the agency hoped, and in 1927 Zon produced a report on forest-flood relationships that essentially recanted the more extreme assertions. At the time, Fernow had observed about the Weeks Law testimony that "it is to be understood that at such meetings considerable buncombe needs to be performed, if you want to handle the half-informed legislators." Perhaps Fernow intended no pun, but Pinchot had begun his meteoric forestry career in Buncombe County, North Carolina.

The Wagon Wheel Gap controversy itself introduces a topic of core importance, that of the independence of Forest Service research from the agency's action arm. In this case, the administrators wanted research to support policy and found the money to do so. The questions then as now are who chooses the research topics and how is funding justified? Are studies that challenge policy more difficult to fund than those that are supportive?

Two scholars have looked at Wagon Wheel Gap and related water studies and suggest that research independence has indeed been hampered by mission-oriented administrators. Political scientist Ashley Schiff and his *Fire and Water: Scientific Heresy in the Forest Service* asserts that the agency's water research, as well as that for the prescribed use of fire, was shaped by institutional priorities. Historian Gordon B. Dodds, biographer of engineer Hiram M. Chittenden who was Pinchot's key opponent in the Weeks Law debate over levies and dams versus forests for flood control, also challenges the objectivity of Forest Service research. The question did not go away with Wagon Wheel Gap, but remains an issue.

PART  RESEARCH STRIVES FOR INDEPENDENCE

The next evolutionary step for Forest Service research was in some ways the most profound of all, the creation in 1915 of the Branch of Research. Secretary of Agriculture David F. Houston saw that Forest Service research was scattered in small and uncoordinated units and directed the chief to take corrective action. Graves of course complied, also explaining that the investigative committees, established three years earlier, had made it more obvious just how essential research independence was. Now the ever-changing administrative priorities would be buffered and at the same time research personnel and efforts could be strengthened. Earle H. Clapp was named chief of re-

Earl H. Clapp served as an assistant district forester in New Mexico before heading the agency's research programs. (Photographed by A. G. Varela; U.S. Forest Service photo.)

search (officially assistant chief in charge), a position he would hold until 1935 when he was appointed associate chief of the Forest Service.

Earle H. Clapp—Research Architect

In the still young agency, it was standard practice to give new appointees their heads in order to show their stuff. Although Clapp received no instructions of substance, he felt strongly that "Henry Graves had a deep interest in research and appreciation of the need for it and its importance." Thus bolstered, Clapp headed west to become acquainted first hand with what was going on at the several experiment stations. He also brought all field researchers into the Washington office for two months to be steeped in research program lore—its objectives, methods, and organization. To learn more about research in general, Graves suggested that Clapp attend the annual meetings of the American Association for the Advancement of Science.

Decades later, Earle Clapp sat in his austere room in Washington, D.C.'s Cosmos Club and drafted his memoirs. In a style that was as frill-free as his surroundings, the former research chief remembered that independence came rather slowly, perhaps a decade would pass before the pieces came together. The most obvious missing pieces were range investigations remaining under grazing administration in both the chief's office and out in the districts. Grazing research had been slated for integration, but associate chief and rangeman Albert Potter, in Graves' absence, had appealed directly and successfully to the secretary to keep it under grazing administration. Too, dendrology under Sudworth reported directly to the chief. The whole of research itself still reported to the Branch of Silviculture and not to the chief; there was a ways to go.

Research independence was of course important, but there were other issues, too. With President Theodore Roosevelt's hand always at the ready to sign a proclamation, the National Forest System had gone through a period of immense growth. Research under whatever label had not been able to even come close to keeping up. At the same time, the instant success of the Forest Products Laboratory had caused appropriations for products research to grow much more rapidly than funding for silviculture and related field sciences. The western experiment stations were staffed by a single investigator and none had advanced degrees. Too, the station leaders were burdened with administrative requirements, such as personally supervising (and even participating in) the construction of research facilities. Clapp's plate was full indeed.

Creation of the Branch of Research was nonetheless a good start. Clapp remembered it this way: "What the reorganization did was to give research a recognition and standing in the Forest Service which it had never had... It helped to break the subordination caused by the enormous, urgent, often highly controversial" effort to bring administration of the national forests up to speed. The self-deprecating architect of Forest Service research also remembered that he had no special training nor a loyal following, "Zon himself told me that he had protested my appointment to Bill Greeley," assistant chief in charge of the Branch of Silviculture. Greeley was a rising star, and by 1920 he would be chief.

World War I

World War I disrupted many American plans, including those of the Forest Service and its Research Branch. Although Graves and many others in the agency joined the military, the chief listed Clapp as an "essential employee of the home government," fairly well assuring that he would not go overseas. Not only were essentially all of research staff assigned to war-related projects but their numbers were increased fivefold to meet the demand for specific information. In the field, research staff sought material that was in short supply, such as walnut for gunstocks and tannin to cure leather. But it was at the Forest Products Laboratory where forestry war work was centered.

When the United States entered the war in April 1917, the Lab had eighty-five employees and a $140,000 budget. A year and a half later, there were nearly four hundred sixty employees, and the budget had grown to $700,000. The Lab made exhaustive studies of spruce for aircraft manufacture and designed crates to protect the endless stream of war supplies headed across the Atlantic. Wheels for artillery pieces and military vehicles alone consumed 120 million board feet of hardwood, species that the Lab had quickly studied. All woods had to be properly cured before use, and kiln technology received special scrutiny in Madison. In all, the Lab drew heavily on its previous seven years experience to bolster the war effort.

In late winter 1917, Chief Graves called key researchers to Washington for a full week of discussions about all aspects of research. When assembled in typewritten form, the final report totalled nearly four hundred pages. As it happened, the chief had to be somewhere else and Associate Chief Potter opened the meeting with informal remarks. Research, Potter asserted, was the reason that the Forest Service was in the Department of Agriculture instead of Interior. Without research, the agency "would be merely an administrative organization."

In addition to presentations on nurseries, fire, seeds, ecology, pathology, economics, and on through the whole list of research needs and activities, there was a lively debate over independence, followed by a straw vote. At that time the experiment stations reported to Clapp on technical matters, but administratively they were responsible to district foresters. Most seemed to feel rather strongly that all reports should go to Washington; district foresters coping with pressing demands would be likely to disrupt ongoing research in order to gain specific information. There is nothing in the report to suggest concern that administrators would in some way attempt to influence research findings to support their actions. Although three-quarters of a century old, much of the report—its contents, thrust, and tone—still holds up today.

Peace returned on the eleventh hour of the eleventh day of the eleventh month of 1918, and only three weeks later on December 5 Chief Graves issued a fourteen-page policy letter on research. He feared that those in research and those with administrative responsibilities did not have similar goals. The administrator too often saw the researcher as distant and detached and thus did not turn to him for needed information. Too, some scientists seemed to be permanently tentative in their conclusions, while the administrator had a job to do that could not wait for yet another round of data gathering and analysis. The chief saw that the perceptions were much worse than the reality and asked that all be more cooperative. To him the greatest gap was the inadequate pool of properly trained scientists from which to draw. Clapp agreed with this analysis and went to work.

Pulling the Research Program Together

Clapp started fitting the pieces of a long term agenda to pull the research program together and to achieve its full independence. Despite the 1915 action, there was still much to do; specifically he wanted to acquire range and dendrology, he wanted to report directly to the chief, he wanted the experiment stations to report to him instead of the district foresters, and he wanted field forest products investigations to be transferred to the regional experiment stations. He also wanted station directors and that of the Lab to attend general Forest Service meetings of the chief and district foresters, in order to better educate both groups on the agency's broader mission.

It was tough going. Clapp needed larger appropriations; instead peacetime programs were cut following the war. Personnel at the Forest Products Laboratory were reduced from 458 to about 300, and research was suspended in California. In 1920 only experiment stations in Arizona, Colorado, and Washington were maintained. It was small comfort that range research, still under grazing administration, suffered the same fate. Also, forest products research had four times the budget as "silvicultural" research, which included all of the experiment stations. It wasn't that products got too much, it was that the stations needed much, much more.

Clapp began to think that a series of enabling laws were needed to address the balance, with little success. Chief Graves took up the cause by advocating a nationwide forest inventory in a speech to lumbermen, who failed to be supportive. As most bills do, those that Clapp had generated quietly died. Research obviously needed to make a stronger case before Congress would do the right thing.

To pull his ideas into a more coherent whole, in 1921 Clapp authored a small pamphlet entitled *Forest Experiment Stations*. In it he proposed creation of ten regional experiment stations, including full representation for the East. He saw some stations with as many as a dozen scientists and each with a budget equal to the current total budget for field research. It was still too soon, and the report did little more than sharpen Clapp's vision.

Even then the Branch of Research was seen as a reservoir of top technical skill. In February 1920 a senate resolution introduced by Arthur Capper directed the secretary of agriculture to report back within three months on the extent of forest depletion, lumber prices, exports, and the degree of forest ownership concentration. The nod went to Research to prepare the report.

Clapp relished the assignment, but the effort made him even more aware of the great need for a forest survey to generate the sort of data that was necessary. Given the shortness of time and the thinness of reliable information, the so-called Capper Report was a good piece of work. However, it got caught up in the intense controversy over regulating forest practices on private lands and in that sense became a political document. The report remains a useful historical compilation.

Meanwhile, there had been some movement toward Clapp's goal. The field experiment stations now reported to him instead of the district forester. Then on July 1, 1921 two eastern experiment stations opened, one in New Orleans with Reginald Forbes as director and the other in Asheville directed by Earl Frothingham. At the same time the Missoula station reopened on a restricted basis. Two years later there were two more stations—at Amherst with Samuel T. Dana in charge and in St. Paul with Raphael Zon as director. Zon had just achieved national prominance through the publication of

Raphael Zon, the first director of the Lake States Experiment Station, worked from this building on the University of Minnesota's campus. (U.S. Forest Service photo.)

Forest Resources of the World in two volumes, which he had coauthored with W. R. Sparhawk ("an outstanding piece of work," according to Clapp), while Dana had returned to the Research Branch after a brief stint as state forester of Maine.

The original experiment stations, such as Fort Valley and Priest Lake, were systematically converted to experimental forests, and most are still actively producing valuable findings. However, as the new generations of stations replaced the original, they did not automatically fare better—budgets remained modest. Philip Wakely remembered that when he started at the Southern Station in 1924, the library consisted primarily of books contributed by staff, and he had to send away for trigonometric tables since the station lacked them. The whole station had only one box of paperclips for all purposes, and thus these stereotypical bureaucratic devices were severely rationed. Wakely was formally allotted twelve paper clips for the year.

Each year would see the Branch of Research grow and also develop depth. The 1924 Clarke-McNary Act authorized a comprehensive study of forest taxation, among many important forestry topics, in cooperation with the states. Also in 1924 another experiment station opened, this one in Portland, Oregon, under Thornton Munger, who had been on site since 1908 with responsibilities for silvics. Yet another indication of growth was the research

One of the Pacific Northwest Station's studies involved seed dissemination. Here, a seed catcher shows the coarse screen top that allowed for seeds to pass through, but kept out birds and rodents. (Scappoose, Oregon, 1925; U.S. Forest Service photo.)

budget for silvicultural studies, which had jumped from $50,000 in 1920 to $285,000 just four years later. The budget ratio to products research was now 2.5:1 instead of 4:1, and the total research budget was $771,000. Looking ahead to 1928, we can see three more experiment stations added in California, Pennsylvania, and Ohio, and a million dollar budget.

The 1920 Capper Report had increased interest in improved utilization. At the time, two-thirds of each tree was lost during either logging or manufacture. The experiment stations looked at depletion from several viewpoints. How to grow timber and how to reforest eighty-one million cutover acres were given high priority, as were economic studies of timber-dependent communities when local supplies were depleted. But there were general issues too; with language that questions the real independence of the Research Branch from the agency's administrative goals. In 1924, the research annual report asked, "Is the lumber industry wise in destroying the productivity of its lands?. . .Is the Nation wise to tolerate this destruction of a capital resource?" Such language helps explain industry's reluctance to go to bat for research in Congress.

Range Research

There was concern that another important resource was being depleted. Since Coville's 1897 report, overgrazing forest range had been a contentious issue, and during the early years, grazing receipts exceeded those from timber sales. To the Forest Service, the simplest solution was to reduce the numbers of stock grazing on national forest ranges. The livestock industry saw the matter differently, and successfully used its substantial political clout to resist. Hard information was needed.

In 1915 the Santa Rita and Jornada experimental ranges were transferred to the Forest Service from the Bureau of Plant Industry. Santa Rita had been established in 1903 and was the nation's oldest experimental range, consisting of fifty-one thousand acres in Arizona. Jornada dated to 1912 and included nearly two-hundred thousand acres of New Mexico. The research goal was two-fold: how to restore, improve, and maintain the basic range resource and how to obtain the greatest returns on livestock. Range surveys, studies of the effects of grazing on water runoff, and the strategic placement of salt to regulate the location of stock were only a few of the investigations. All this was brought together in 1919 with publication of USDA Bulletin 790, *Range Management on the National Forests* by James T. Jardine and Mark Anderson. The book quickly became the range manager's bible.

James T. Jardine took this photo to show the heavily grazed range around the Slease Goat Ranch on Trujilla Creek, Gila National Forest, New Mexico. (U.S. Forest Service photo, 1916.)

By the 1920s, the range situation became very political through the direct involvement of Congress, but Forest Service studies effectively supported the gradual reduction of livestock. The parallel issue of how best to determine the value of the forage and whether the permittee's fee should be based upon the value of the resource or the lower cost of administration were pieces of the range puzzle that would remain controversial. In recent times the substantial increase of administrative costs due to the adoption of environmental safeguards has in many areas made these costs greater than the market value of the resource, leading to charges of below-cost grazing permits. The line between science and policy is ever moving.

Research Natural Areas

The Forest Service gives official credit to the Santa Catalina Research Natural Area in Arizona's Coronado National Forest for being the first, established in 1927. Even so and granting that other regions can also quibble about the definition of "first," in 1926 at Washington's Wind River Experimental Forest, 280 acres were set aside and withdrawn from all disruptive use and occupancy. Station director Munger wrote of the "need to preserve intact examples of the principal virgin forest types" through such setasides.

Indeed, internal debates over establishment of wilderness areas had for several years previous incorporated a similar scientific rationale. There was a scientific value to wilderness in that each area would provide an ecological benchmark against which to measure the impact of human intervention on managed lands. While wilderness areas might measure as much as one million acres, however, research natural areas are on the average of a thousand acres, ranging from forty to ten thousand acres. The 250th RNA was established in 1992, with a rather broader definition—permanent protection to maintain biological diversity, conduct nonmanipulative research and monitoring, and to foster education—that goes much beyond Munger's "virgin forest types."

Research Councils

Minutes for chief and staff for January 22, 1925 show that Clapp reported on the meeting of the Lake States Forest Research Council, which had convened in Chicago on the heels of that of the American Forestry Association. He announced that the Northeastern Council would meet in New Haven on February 7, and the newly formed Appalachian Council on February 13 in Asheville. A southern council was in the process of formation. A year later, Northeastern Station director Dana stated that the Northeastern Council had been meeting twice a year since 1924 with very satisfactory results. The council was especially concerned about gypsy moth infestations. Too, the council proposed giving emphasis to spruce and hardwood research, as scientists at Harvard Forest and the Yale School of Forestry were taking care of white pine.

McSweeney-McNary Act

In the history of Forest Service research, the year 1928 is a turning point—to call it a watershed would be an exaggeration—that merits extra attention. This is the year that the last of the "Hoot Mon" laws got on the books. The first was the 1924 Clarke-McNary Act that we have already seen, then in 1928 the McNary-Woodruff Act greatly expanded land purchase under the 1911 Weeks Act, followed a month later by the McSweeney-McNary Act that implemented Earle Clapp's research program, including a nationwide forest survey. We need to go back four years.

On April 30, 1924 Chief Greeley addressed the National Academy of Sciences. He recounted that an academy committee in 1897 had provided the nation a yeoman service by recommending vast areas to be added to the national forest system. These additions had been completed and more, but there were still forestry problems to be faced, such as timber depletion and millions of acres of cutover lands that were neither farms nor forests. Needed, according to Greeley, was an "evolution in the land practice and forest industries of this country." The evolution would prove inadequate "unless a comprehensive scientific foundation can be provided for it." He called for advances in silviculture, economics, products, and a nationwide survey of forest land to provide a reliable inventory. The chief asked the Academy to study the forestry research situation and make recommendations.

Shortly thereafter, Greeley called Clapp to his office, telling him that the National Academy of Sciences had convened a committee to study forestry research and report back. Clapp remembered that he was caught by surprise and on the spur of the moment asked whether he could take on the job. Greeley said no, and offered no explanation. However, it seems safe to assume that Greeley wanted the study to be independent of the Forest Service and for that reason declined to allow Clapp to participate.

The Academy responded favorably to Greeley's request, appointing a distinguished committee that included former chief Henry Graves, then dean of forestry at Yale University. The Society of American Foresters helped gain funding from the Carnegie Corporation to allow for visiting research institutions across North America and Europe. The committee report on federal, state, university, and private forestry research was published in 1929. However, as with the 1897 Academy report, by the time the official document was filed, significant actions had already been taken. Whatever its direct influence, the report provides a broad blueprint of forestry research goals and an assessment of research institutions.

In November 1924 Greeley had also presented a variation of his NAS speech to the Society of American Foresters. This time Clapp would be involved, as the SAF appointed him to its committee to respond to the chief's charge. Thus, there were two committees in place studying forestry research, and there is evidence that they exchanged information. The SAF committee reported earlier—in 1926—and in substantially greater detail than the NAS committee. Also, the SAF effort zeroed in on Clapp's idea for a national program, including language for a federal research organic act.

Harry Irion of the Forest Service legal staff drafted a research bill for congressional consideration. The main snag was resistance from the bureaus of Entomology and Plant Industry. As did the Weather Bureau in the Department of Commerce, these two Department of Agriculture agencies had cooperated with the Forest Service for years to study mutually important topics. Now, however, the two bureaus did not want to be included in a Forest

Service bill. Secretary William M. Jardine intervened, calling a meeting in his office that included Greeley and Clapp. The secretary pulled rank, overruling Entomology and Plant Industry because the bill was designed to carry out a desirable program. Now to get it introduced.

Oregon Senator Charles McNary, cosponsor of two previous forestry bills, was an obvious choice for the Senate side, and Wilson Compton, head of the National Lumber Manufacturers Association, recommended Congressman John McSweeney from his home district in Ohio for the House side. This latter recommendation was very significant, because it showed at long last open support from the forest industry for Forest Service research, other than that in products. A bit earlier, Compton had written to Weyerhaeuser's George Long, assuring him that the industry was going to press for passage.

One last knot to tie was securing approval of the Budget Bureau and President Coolidge himself. Clapp knew that Coolidge looked to the state forester of Massachusetts for forestry advice, and so the research chief made certain that his state colleague was well informed and supportive. The president approved the measure but stipulated that there would be no appropriation increase until the following fiscal year. Ducks neatly in a row, the bill was introduced—hearings showed only support—and was passed into law on May 22, 1928. Secretary Jardine sent Clapp the pen that the president had used to sign the law, saying that it was "a splendid achievement."

Clapp had indeed labored eight years for some sort of enabling legislation to pull together the piecemeal, station by station approach. He kept a steady stream of letters aimed at experiment station directors and others who might lobby with their congressional delegations. His memoirs suggest that for him it was a lonely battle with no open support from Chief Greeley beyond those two 1924 speeches to the National Academy of Sciences and the Society of American Foresters.

To Clapp the McSweeney-McNary Act accomplished several important things: it gave the Branch of Research a broader place at the Forest Service table, it strengthened the basis for dealing with other agencies, it dealt with nonfederal research needs, and it began to balance silvicultural and products research. One of the most important points, of course, was authorization of the Forest Survey, the national inventory that Clapp had long felt was essential. The law was so broadly based that the Forest Service would not seek additional research legislation until 1978. Because the Forest Service already had authority to experiment and continue investigations and had ongoing appropriations for research, the McSweeney-McNary Act was not so much a matter of granting new authorities than it was one of specificity.

A New Deal for Research

No one had said that it would be easy. Clapp's long-sought McSweeney-McNary Act was scarcely a year and one-half old when a plunging stock market rudely announced a depression that had been lurking just over the horizon for half a decade. Of course, since no one realized that the depression would last until December 7, 1941, reactions were often stop-gap and piecemeal, as the Forest Service became heavily involved in various depression-related programs.

Nonetheless, the research law took hold, expanding programs and increasing appropriations. President Coolidge had extracted a pledge that no increases would happen for a year, so it was the 1929–1930 fiscal year that saw the first McSweeney-McNary appropriation increases—20 percent to $1.2 million and a new station authorized for Ogden, Utah. Tucked away in the budget was $40,000 to begin the Forest Survey and also money to study fire insurance for forests in the Pacific Northwest and private forestry in the South. The following year, there would be money to study soil erosion.

The appropriation for silvicultural research was now a full 70 percent of that for products. Adding range and taxation studies to those for silviculture, products accounted for one-half of the total research budget. Two years later products was 77 percent of the silviculture budget and only 39 percent of the total. The products appropriation had remained constant, while other research categories increased. During the same period, Congress also appropriated nine hundred thousand dollars to build a new Forest Products Laboratory in Madison, as the original structure had become woefully inadequate. In his memoirs, Clapp tracked each year's budget; he ought to have been pleased that congressional priorities and his own were now closely aligned.

Copeland Report

As with the earlier Capper Report, the Branch of Research was largely responsible for the 1933 Copeland Report, officially named *A National Plan for American Forestry*. Published in two volumes, the 1677-page study delved into all aspects of forestry, in some ways similar to today's Resources Planning Act Assessments and Programs. Clapp headed the effort, and his personal papers held by the National Archives contain many anecdotes and insights.

On March 30, 1932 the Senate approved Senator Royal Copeland's resolution 175, calling for what became a massive Forest Service effort. In July of that year, Clapp wrote to all experiment station directors, saying that the

The Forest Products Laboratory located in Madison, Wisconsin began research-ing wood and wood products in 1910. This structure was constructed in 1932, during the New Deal. Today, the building is on the National Register of Historic Places. (U.S. Forest Service photo.)

Copeland resolution offered a "great opportunity" to restate American for-estry in a positive fashion. Although a magnificent effort that reflects the state of American forestry art as of 1933, the Copeland Report landed with a dull thud, companion bills for implementing its myriad recommendations dying quietly in committee. A decade later, Clapp wrote bitterly about lack of strong leadership from the chief, and lack of interest by President Franklin D. Roosevelt—assessments that probably tell us more about Clapp's state of mind than the true situation. Most likely a major reason for failure was strong and repeated advocacy for federal regulation of the forest industry, a concept that Congress would not support.

The report may have failed to achieve the agency's enforcement goals, but it was not a failure. It had taken its toll, however; on a personal and bit of a sour note, southern silvicultural scientist Philip Wakely who had been as-signed to the study remembered it as a two-year disruption and commented that "we were so sick and tired of it that most of us never looked at it."

Nonetheless, the report brought together the latest forestry information, of great value to all sectors. For example, forestry students often found them-selves using it as a required textbook. Also, the report included two sections on Forest Service Research, a history and appraisal by Clapp, and a recom-mended research program by Earl Frothingham, director of the Appalachian Forest Experiment Station in Asheville, North Carolina.

Clapp was bluntly candid as he portrayed the evolution of the Branch of Research as a nearly heroic struggle for independence from the administrative side of the agency. Some researchers might have felt uncomfortable when Clapp wrote, conceivably about them, "That the national requirements for forest research have not been met during the last decade is primarily because sufficient men with the necessary mental equipment and training have for one reason or another been nonavailable." Clapp also pointed out that the agency had routinely used Research as a dumping ground for those who did not fit well in Administration. The good news was, despite the problems there were many research accomplishments to point to.

Clapp's fourteen-page chapter contains an impressive list of accomplishments: volume tables for fifty tree species, silvicultural treatments showed promise for reducing forest fire damage, range studies had classified the most important forage species, forest and range influence on water flow was better understood, utilization studies improved timber harvest methods, new lumber grading methods yielded more value from each tree, and proper drying increased wood durability. Clapp concluded his summary of accomplishments with a report on forest economics—data on lumber production and consumption was being collected in conjunction with the Bureau of the Census, and the Forest Survey had begun in earnest in 1930.

Frothingham's proposed research program pointed to a condition that is much less of an issue today. Although Forest Service research had been coordinated and given independence, there was much forestry research conducted outside of the agency: forest pathology was in the Bureau of Plant Industry, forest entomology was in the Bureau of Entomology, the Bureau of Chemistry and Soils handled naval stores, and the Biological Survey looked to forest biology. finally, the Bureau of Fisheries was responsible for research on fish living in forest waters, and the Weather Bureau was a major contributor to fire research. In subsequent years, expansion of the Forest Service research mission and government reorganization would make the agency less dependent upon the work of others. Of course, collaboration is and was central to much research, and thus the federal effort has not moved toward isolation.

Forestland Taxes

While the Copeland Report looked broadly at forestry issues of the day, three other research studies were very specific and dealt with issues that were controversial in the 1930s and are still debated today. Published in October 1935, *Forest Taxation in the United States* is a 681-page work that was designed to put significant forest tax questions to rest. The 1924 Clarke-McNary Act had authorized the effort, which was headed by Yale professor Fred Rogers Fairchild, a nationally prominent tax expert who had published on forest taxation as early as 1908. Unfortunately, rigid adherence to a departmental publication rule that "studies" could not be indexed, only "books" could, caused a meticulously prepared index to be jettisoned. Thus, the only road map to this hefty contribution is a minimal table of contents for each "part," as regulations allowed "chapters" to appear only in "books."

As an accommodation to Fairchild, the tax study was headquartered in New Haven. He assembled a highly talented crew, including R. Clifford Hall. Fairchild's Yale responsibilities were heavy, and so Hall managed the tax program on a day-to-day basis. One of the key questions was the impact of property taxes on timber liquidation rates. In 1932 President Hoover's Timber Conservation Board asserted that "The present and prospective annual burden of taxation on mature standing timber is the most important single present factor forcing the sale or cutting of timber without due regard to the current market demand for forest products."

Not so, said Fairchild and Hall, there was no evidence to support such an allegation. However, debt burden did cause much timber to be logged, even into a glutted market with prices lower than the cost of production. The tax experts reasoned that industry's loudly voiced position was mainly a defensive reaction to the vigorous Forest Service campaign to attain authority over private logging practices by claiming serious overcutting. In this scenario, high taxes that could be blamed on government were driving the overcutting; it was a convenient argument, but in private lumbermen had since 1908 reported to Fairchild that property taxes were not crucial to their decisions on scheduling logging. Clearly, the research finding was not what foresters and lumbermen wanted to hear, and the taxes-drive-logging story would continue to be taught in the nation's forestry schools for at least another generation. As noted above, tax debates continue today and continue to mix folklore with fact.

Douglas-Fir Study

In the Pacific Northwest, politics and science were mingled during an investigation of alternatives to clearcutting old-growth Douglas-fir. Axel Brandstrom worked at the experiment station in Portland, and Burt T. Kirkland was on the forestry faculty at the University of Washington. The two collaborated on finding a way to make lumbering more profitable—or profitable at all—during the depths of the Great Depression. They proposed in a 1936 published report that selectively logging Douglas-fir would both increase profits and also be silviculturally acceptable. Station director Thornton Munger had opposed publication, but Raymond E. Marsh, Clapp's assistant in the Washington Office, overruled. The controversy was softened a bit by asking the Charles Lathrop Pack Foundation to publish, instead of the agency doing so.

Not only Munger opposed publication based upon its silvicultural deficiencies, but so did timber management specialist Walter Lund of Portland's regional office. However, Regional Forester Charles Buck became a quick advocate of the Brandstrom-Kirkland recommendation, mandating that clearcutting be replaced with selective logging. The edict remained in effect for a half-dozen years, when more carefully designed clearcuts became the norm.

In 1956 Leo Isaac, a Portland silvicultural specialist who had always believed that selective logging in old growth Douglas-fir to be scientifically unsound, published a long term study that pointed to the silvicultural deficiencies of Brandstrom and Kirkland's two-decade earlier work. But the controversy was not over; in 1969 Sierra Club forester Gordon Robinson published "Excellent Forestry," an anti-clearcutting article that drew heavily upon Brandstrom and Kirkland. The arguments continue, echoing the half-century debate over the proper management of the redubbed "ancient forests" in the Pacific Northwest.

Range Controversy

Lest the reader still believe that research—at least Forest Service research—exists in some sort of ivory tower, there is yet another controversial report to examine, this time about the condition of the American range. A Forest Service-livestock industry detente of the 1920s was wearing thin; new information was needed.

Years later, Clarence Forsling, a Forest Service range specialist who would be named head of Forest Service research, remembered being in Europe on official business. He received a cable from Clapp, ordering him home at first opportunity. He returned to help finish *The Western Range*, published in 1936 but begun in 1932. Clapp was especially proud of the finished product, seeing it as a companion to the Copeland Report. The research chief would transfer one hundred congratulatory letters to his personal files.

The six-hundred-page report was blunt; rangelands were seriously deteriorated for two basic reasons—the Department of the Interior had failed to live up to its management responsibilities, and also because the 1934 Taylor Grazing Act allowed the livestock industry too much autonomy. Not incidentally, the industry had to deal with two federal agencies; the logical solution was to have the Forest Service manage all federal rangelands. Predictably Secretary of the Interior Harold Ickes and the livestock industry saw the situation differently—the word vehement comes to mind—and vigorously challenged the Forest Service research findings. Stockmen countered with a study of their own; overgrazing was not the villain, it was too little rainfall that caused range depletion. As so often happens, the interagency confrontation turned to stalemate. Fences were apparently mended a bit in 1944 when Secretary Ickes asked Forsling to head the department's Grazing Service. If the Forest Service could not be in charge of all federal grazing, at least its people would run it. For the record, federal range management remains controversial, and there is wide acceptance that public range lands are not in very good shape. Too, the growth of range research has not kept pace with other subjects.

Statistical Training

Francis X. Schumaker moved to the Washington Office of the Forest Service in 1930 from a forestry professorship at Berkeley. One of his former students, Verne L. Harper, was in 1931 his first trainee in advanced statistical methods. Harper spent a year in Washington furthering his education in mathematics at Georgetown University and in statistical methods at the USDA Graduate School. With Schumaker's guidance, Harper statistically analyzed naval stores data that he had brought with him from Starke, Florida. These analyses would become a teaching tool in Schumaker's yearly classes for selected researchers, and they became a useful reference for entomologists who found that the time of day of gum yields was correlated with active attacks of the southern pine beetle. In 1937 Harper and Lenthall Wyman used the data to author *Variations in Naval Stores Associated with Specific Days Between Chippings*, USDA-Forest Service Technical Bulletin No. 510.

In August 1936 the Forest Service held a two-week statistical methods seminar at the Southeastern Experiment Station in Asheville. Attending were Forest Service scientists and forestry professors. The seminar was aimed specifically at forestry issues, as opposed to the broader, three-month statistical methods course offered by W. Edwards Deming at the USDA Graduate School. Leading the Asheville sessions was R.A. Fisher, an eminent British mathematician whose statistics textbook served generations of students. His focus on practical problems was stimulating. At another time, Fisher had warned that "to consult the statistician after an experiment is finished is often merely to ask him to conduct a post mortem examination. He can perhaps say what the experiment died of." The agency heeded the warning and demanded the increased use of statistically valid experimental design.

George Jemison, who three decades later would succeed Verne Harper as deputy chief for research, was a forestry student at the beginning of the 1930s. He took a "poorly taught" course in statistical methods. Each subsequent year, however, as more and more forestry students graduated with familiarity with statistical design, they created an atmosphere that eased Forest Service adoption, at least in principle. Eventually, Jemison would teach a statistical training course at the Rocky Mountain Experiment Station; at the Southern Station, Roy Chapman, who had tutored for several years under Schumaker, influenced the adoption of statistical design for new projects. The other stations tell similar stories, as research moved ever more in the analytical arena.

In 1938 the National Research Council published its study of Forest Service research. In sum it tabulated the various studies and provided narrative descriptions. We can see, for example, that of the 1,308 total projects, 8 percent or 102 dealt with ecology, 28 percent or 364 were silvicultural in nature, and 10 percent or 129 treated products, and so on through the dozen fields of study that comprised Forest Service research. It also laid out the scientific standards of the day: in 1938 experimental design was to yield "statistically valid results."

Anecdotes abound on data collection and analysis; until methodology and instrumentation were standardized, statistical applications would be limited. Robert Cowlin, who would eventually become station director in Portland, was assigned to the Forest Survey in the Douglas-fir region. His special assignment was to test a new measurement system called the line-plot method. He later calculated that 960 man-days were required to complete the test, days that were typically much longer than eight hours. Schumaker was an interested observer, and his considerable skill with poker served him well playing with the survey crew after hours. Cowlin remembered that they got their money back by betting on tree diameters, which were confirmed by careful measurements. As it turned out, "Schu had a tendency to under estimate the large old-growth Douglas-fir."

Harry T. Gisborne, the most eminent of the early fire scientists, had come up with an idea for an inexpensive instrument to measure windspeed, a crucial factor in predicting rate of fire spread. A local plumber hand-made one hundred and sixty of the devices, meaning that each was a bit different and had to be individually calibrated. One at a time, George Jemison mounted the gauges on the front of his car. As his wife drove at five, ten, and fifteen miles per hour, he lay on the fender and counted revolutions. Jemison recounted that "it was primitive but very effective compared with other methods." In just this way, researchers with imaginative minds and clever hands would translate technical problems into usable forms for those on the ground to apply. In recent times, the process is called technology transfer.

While Jemison was counting revolutions, farther west in Oregon, Leo Isaac was trying to figure out just how far Douglas-fir seeds would be carried by the wind. This was not an abstract issue; to achieve satisfactory natural regeneration after clearcutting, the cutting unit could not exceed the distance that seed from cones high up in the tree can fly in the prevailing wind. In the tradtion of Gisborne, Isaac attached a paper Quaker Oats package to a kite and flew it over a snow-covered field. A jerk on the string tied to the lid allowed Douglas-fir seed to disperse. The snow made it easy to trace the pattern, and from repeated tests at various windspeeds, Isaac developed a standard for the optimum-sized clearcut.

Other Depression Research

The larger context of the time, however, was the Great Depression. The Forest Products Laboratory responded by emphasizing creation of new products. After all, the forest industry was engaging in a holding action; for each ten thousand employees only three were assigned to research activities. This number did not compare well with the chemical industry, for example, which could boast of 303 in research for every ten thousand.

Obviously, the private forestry sector was developing few new products, and the Lab attempted to fill the void. The first laminated beam appeared in Madison in 1935 and the first prefabricated house in 1937. The Lab also studied lignin, one of wood's primary constituents, and products that could be chemically derived from wood. The Lab published the *Wood Handbook* in 1935, an invaluable reference work; when it was revised in 1955, the Government Printing Office reported that it was among the top twenty-five government publications.

The Lab received national prominence during the widely followed Lindbergh kidnapping trial. Crucial testimony leading to conviction came from wood scientist Arthur Koehler, who positively showed that a member in the ladder used in the kidnapping had been cut from the attic floor of the defendant's home. A few months later, Koehler was called upon to similarly identify wooden shards remaining from a cigar box bomb used to murder three people. The favorable publicity assured that more and more often the Lab would be asked to identify wood, and not just for criminal proceedings.

The forest industry supported the Lab and its budget, giving it full credit for the "remarkable expansion" of the industry in the South, for bringing plywood beyond being "crude and expensive," and for its "almost revolutionary" advances in wood seasoning and preservation. Leadership in getting facts, proving their effectiveness, and making them available to the industry was the Lab's "basic job."

Out in the field, there was also much activity. In 1935 the Rocky Mountain Experiment Station opened its doors in Fort Collins, Colorado, and in 1939 a modest program began in Puerto Rico. Construction of research facilities in general got a big boost from Civilian Conservation Corps labor. The CCC applied treatments such as shelterbelt experiments, thinned plots used in growth studies, and constructed access roads to research areas.

In 1936 the Oxford Paper Company offered thousands of poplar hybrids that it had been studying for over a decade to the Northeastern Experiment Station. A year later Seattle lumberman James G. Eddy donated the Eddy Tree Breeding Station in Placerville, California, to the Forest Service; both gifts assured that forest genetics would be a significant research activity.

There was growing interest in forest cooperatives, such as those long used by farmers for marketing their products. In 1936 the Otsego Forest Products Cooperative was established at Cooperstown, New York, and several hundred woodland owners joined. The cooperative built a sawmill, developed forest management plans for members, and sought new markets. The Northeastern Station assisted in studying cooperative organization, in developing an inventory method for standing timber, investigating silvicultural systems, and devising a log-grading system. Other less elaborate forest cooperatives devel-

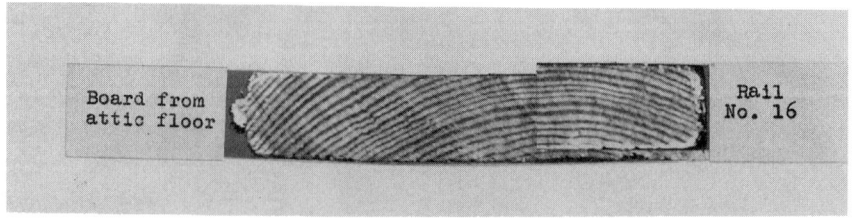

Wood identification by scientist Arthur Koehler used as evidence in the Lindbergh kidnapping trial. (U.S. Forest Service photo.)

oped about the same time in the Lake States. All had a common objective, that of assisting rural people to find markets for their timber.

The Norris-Doxey Cooperative Farm Forestry Act and the Bankhead-Jones Farm Tenant Act, both enacted in 1937, supported economic studies related to farm forestry and agricultural cooperatives. These two acts bolstered the various forest economics research programs. At the Lake States Station, director Zon reported that 17.3 percent of his budget supported economic studies, to which he included recreation research. In his view, resorts received too much recreational attention, and emphasis should be shifted to include other forms of outdoor enjoyment.

Artificial Regeneration in the Southern Pine Region, USDA Technical Bulletin 492 by Philip Wakely, appeared in 1936—in time to become the Civilian Conservation Corps bible for the millions upon millions of seedlings it was planting across the South. It was his first major publication and had several printings.

And yet another scientist was at work, developing a process that would make the field forester more effective. With the Bureau of Entomology and Plant Quarantine but on detail to the Forest Service, Paul Keen carefully studied ponderosa pine bark beetles, a serious predator of the oldest and commercially most valuable individuals. Eventually the Keen Classification System evolved; by comparing simple drawings of a range of ponderosa pine crown conditions with those in the forest, the field forester could accurately predict the likelihood that a particular tree would soon succumb to beetle attack. Using this information in conjunction with other factors, the forester then would mark the trees to be selectively logged. In just these sorts of ways, scientists added bits and pieces to the forester's tool kit.

Franklin B. Hough in 1882 had advocated experiment stations in all regions to capture the diversity of the American forest. A half-century later, this had been accomplished through creation of the basic experiment station network, and with the conversion of the original stations into experimental forests. Also, the establishment of research natural areas and the introduction of statistical design and problem analysis were new threads in a still young research tapestry.

Some who were active at the time felt that the single most valuable program was the establishment of growth plots, species by species. It would be the next generation of scientists who reaped the benefits of having access to continuous data collected over decades. Much of this information is critical to current understanding of forests. A related effort was the establishment of additional experimental forests and ranges; eventually there would be more than eighty.

The Forest Products Laboratory rounded out the picture by expanding the breadth and efficiency of use. Finally, the scientists themselves, only sampled above, provided technical and professional leadership within the agency and throughout the world.

Clapp Era Ends

In 1935 Chief Ferdinand Silcox asked Clapp to be associate chief. Candidates for his successor as head of the Research Branch included Aldo Leopold, Forsling, and Dana. However, Raymond E. Marsh, Clapp's longtime assistant, was selected to be "acting" head of research. Clapp's many reservations about Marsh's ability to handle the job seem to have been borne out, because in 1937 Forsling and not Marsh was named assistant chief of the Forest Service in charge of research.

Typical of his dour temperament, Clapp wrote to Forest Service scientists across the country saying that he was "leaving Research with an uneasy feeling about its future." Still refighting old battles, he reminded his colleagues that "the Forest Service in general has been indifferent to or has actively opposed practically every constructive move to develop research." Sadly, the man who had done so much saw his efforts as the "most thankless and difficult duties." He went on to urge ever higher standards, so that he could "leave Research with a greater feeling of assurance of its future." From the written record it is difficult to know just what would have had to happen for Clapp to say that he was pleased, but the direction that Forest Service research has taken during the past half-century ought to have satisfied him—much of his imprint remains.

When Chief Silcox was felled by a fatal heart attack in 1939, Associate Chief Clapp was named acting chief. However, a president angered by agency efforts to prevent transfer of the Forest Service to Interior in turn prevented the ex-researcher from becoming chief in his own right. Lyle Watts was appointed chief of the Forest Service in 1943, and Clapp retired shortly thereafter. Harper remembers that Clapp remained interested in the Research Branch, and that they became friends. In private, Clapp "smiled and related anecdotes, not the usual dour persona he projected."

Clarence Forsling and World War II

Clapp was still associate chief in 1937 when he wrote to Forsling, asking him to be chief of research. The request is not surprising, as a year earlier Clapp had appraised Forsling as "one of the strongest and most promising of the forest experiment station directors." His 1931 publication that treated the effect of grazing on water runoff and soil erosion (USDA Tech. Bull. 220) had brought him wide respect. But Forsling was not certain that he wanted the job and queried Clapp about certain aspects of it. Typically, Clapp was not deterred by a few questions; as Forsling later remembered, when he "made up his mind about something, he just went ahead unless he ran into a stone wall." In the end, Forsling left the experiment station directorship in Asheville, where he was succeeded by future chief Richard E. McArdle, to become assistant chief for research.

As had Clapp, Forsling worked hard to obtain larger appropriations. He also saw that consolidation and workload redistribution was needed in some of the field operations. Too, he intended to expand studies dealing with range, forest economics, forest products, and the effect of plant cover on water runoff, the latter topic of special interest to him. He started on this agenda, but World War II changed everyone's plans.

Forsling was returning by train to Washington from a conference at the Priest River Experimental Forest in Idaho. During a switching he overheard trainmen talking about the "Pearl Harbor incident." Civilian budgets were promptly slashed, but supplemented by requests from the War Department for information. Many Forest Service scientists were either called into the service or transferred to newly created war agencies; by 1945, 198 scientists were in the armed forces, with 2 killed in "the line of duty." Field work was halted, and those remaining at a station were mostly engaged in work connected with the war. Some of the stations, especially the Southern, became heavily involved with field surveys and reports required by the War Production Board. There were few exotic assignments, and some scientists found themselves in the tropics seeking supplies of the ultra buoyant balsa for making life rafts, quinine to treat malaria contacted by troops in tropical zones, and rubber substitutes following loss of imported supplies due to Japanese military intervention.

New rubber sources to be explored were the domestic cultivation of a Russian dandelion called kok-saghyz and guayule, a southwestern shrub. Forest Service researchers provided technical assistance in Minnesota and Michigan for kok-saghyz and in California for guayule. By war's end the supervisor of the Chippewa National Forest was using kok-saghyz tires on his official car, and guayule sheets were formed into leakproof fuel tanks for military aircraft. The guayule process was more productive than kok-saghyz, yielding three million pounds of rubber before foreign supply was once again available following the war. The development of synthetic rubber from petroleum had occured early enough in the war to keep guayule to a pilot project level. This wartime research was revisited during the 1980s when there was interest in developing arid land crops and bioenergy.

Scientists also examined tropical woods as potential bridge timbers for the crash program to complete the Transamerican Highway to South America. Others, including Forsling, Harper, and future deputy research chief M. B. Dickerman, were assigned to the War Production Board. The board aimed to assure a timely supply of forest products to the war effort by reducing bottlenecks, clarifying priorities, and retaining realistic prices for both military and civilian purchase. To some extent, funds for war-related assignments made up for a reduction in civilian research programs.

But the real wartime star was the Forest Products Laboratory in Madison, Wisconsin. Low-temperature glues for laminating beams, preservatives for wood used in the tropics, improved hardboard production, and enhanced wood alcohol distillation processes (used for rubber substitutes)—twenty gallons per ton of wood—were some of the laboratory's contributions. It was the revolution in packaging technology, however, where the laboratory was truly center stage. The effort included the training of some fourteen thousand personnel between 1942 and 1945 in the efficient use of packaging materials, economy of space, and protection of contents. Now, each radar, each case of ammunition, each truck engine, and on down the seemingly endless list of military supplies shipped across the Atlantic and Pacific oceans more and more often arrived in workable condition. Also, the specially designed container was available to hold a disabled unit as it was returned for repair. Much of the Allies' success during the war was due to America's vast industrial capacity, but damaged equipment won no battles. Customized boxes and crates is a fitting theme for a chapter on World War II, as it was for the Great War of a generation earlier.

International Forestry

Creation of the United Nations, the Marshall Plan to assist rebuilding of war-torn nations, and similar activities assured that the United States would remain active around the world following the war. Given the fundamental importance of wood for fuel and for building, it is no surprise that the Forest Service was called upon to assist the Food and Agriculture Organization (FAO) that had established a Branch of Forestry and Forest Products. Chief Lyle Watts was appointed to an FAO advisory committee as plans went forward to "set up international forestry statistical services, assist governments with advice on forest policy, send out missions to make scientific studies, promote research and circulate findings among nations, and facilitate exchange of scientific personnel."

Forsling later recounted just how forestry came to be included in FAO. Watts had turned international activities over to the Research Branch for implementation. Egon Glesinger, secretary to an international lumber sales association headquartered in central Europe, offered to assist Forsling, and together they got Lester B. Pearson, later to be prime minister of Canada, involved. Now the time came that the United States needed to decide whether it would join a permanent forestry organization in FAO, and Forsling drafted a letter to Secretary of State Dean Acheson for Secretary of Agriculture Claude Wickard's signature. Undersecretary of Agriculture Paul H. Appleby saw the draft first, sending it back to Forsling saying that "the Forest Service should not get into things like that, and that the United States should not become involved in forestry all over the world." Forsling did not give up and was able to persuade Appleby to informally ask Acheson directly. The secretary of state replied, "By all means the Department of Agriculture should be involved in that phase of the FAO program." Former Forest Service chief and Yale's forestry dean Henry Graves was drafted to organize the permanent FAO forestry committee.

There were other international vehicles, such as the International Union of Forestry Research Organizations (IUFRO) and later the U.S. Agency for International Development (USAID) and the World Bank, that would carry the American forestry ball overseas. But except for a small research outpost in Puerto Rico since 1939 that included the inauguration of *Caribbean Forester*, a journal that would be published for twenty-two years, it would be a decade or more before such international activities became more than an ad hoc Forest Service effort.

The Kotok Years and Postwar Research

Forsling moved over to the Department of the Interior to be chief of the Grazing Service (merged with the General Land Office in 1946 to be the Bureau of Land Management), and Edward I. Kotok was Watt's choice to be assistant chief for research. It would be Kotok's challenge to deal with research needs during the chaotic postwar years of reemploying veterans, a booming housing industry, and a general sense of affluence and optimism across America.

Postwar was a time of growth. The National Housing Authority called for nearly three million new homes by the end of 1947, and the Forest Products Laboratory gave priority to kiln-drying research so that properly cured lumber would be available for new construction. All facets of Forest Service research would be affected by the increased tempo; Kotok testified to the House Appropriations Committee in 1946 about the need for additional research for timber, range, and water resources. He described twelve experiment stations and asked for two more, one in Alaska and one in the Great Plains. All this was part of a five-year research plan to assure "orderly development." Much needed were new field facilities, but Kotok assured the committee that 80 percent of new money would be for personnel, and the balance for "incidentals" including sixteen new cars. The committee added a half million dollars to push the total research appropriation over five million.

Next year Congress approved nineteen new research centers for a total of fifty-three. Kotok felt that about thirty more centers were needed before all important forest types and economic units were covered nationwide. The idea for research centers—away from station headquarters—came out of the South. Hardwood research at Stoneville, Mississippi, naval stores studies at Lake City, Florida, and forest management investigation at Athens, Georgia, were prime examples of subject-specific sites, and not incidentally with strong congressional support.

As Kotok's testimony shows, each center was specifically identified in the appropriations measure, as the concept spread nationwide. By providing focus to funding needs, and related congressional involvement, the centers greatly helped bring research together during the postwar shift to civilian priorities.

A decade later, some felt that the centers began to get out of hand, with strong center leaders and strong local ties tending to lessen allegiance to the stations' broader agenda. Too, there was unwanted pressure from Congress for unneeded centers. Future leadership would be challenged to bring control without losing creativity.

There were specific advances during this period. For example, in the Pacific Northwest, researchers looked at ways to ease the transition that was just underway between old growth timber stands and their second growth replacements. Partial cutting systems in the South and Rocky Mountain forests showed promise, but not so for the Far West. Research continued on the use of aircraft for seeding cutover lands, and hybrid seedlings produced at the Institute of Forest Genetics were showing great potential as heartier and faster-growing nursery stock. By 1949 the scientists were testing "new tools" for the forester—the powerful insecticide and herbicides DDT, 2,4-D, and 2,4,5-T.

J. Alfred Hall was a Ph.D. chemist and had held a variety of research assignments since he joined the Forest Service in 1930. During World War II he became the Principal Biochemist, essentially trouble-shooting a variety of war-related forestry research, such as the guayule program in California. As he roved he developed the concept of the Forest Utilization Service, a national network that would help the private sector achieve a greater degree of utilization. In 1945 he was named director of the Pacific Northwest Forest and Range Experiment Station, and with Kotok's concurrence one of his first actions was to establish an FUS unit in Portland. The following year there was an FUS unit at the Central States Experiment Station, and other stations were soon to follow. All such units carefully coordinated their efforts with the Forest Products Laboratory; Hall would become director of the Lab in 1951.

The 1949 annual report of the chief contained a new wrinkle; research activities led off followed by State and Private Forestry and the National Forest System. The report's theme was "New Knowledge in Forestry"; Chief Watts introduced the report by saying that "through well-organized research we can learn how to increase forest productivity and achieve more effective conservation. Research can show us better ways of doing things, at lower costs." Kotok's section was divided into seven topics: forest management, fire, range, forest influences (a term that would soon be phased out and be generally replaced with watershed), economics, products, and tropics. He ended with a look to a new research program, as there was always more to learn, and the need for improved application of research findings. "To find the right answer to a problem is only part of the job. The problem will be met only when the new knowledge is put into practice." Results, the chief researcher admonished, "must be brought out of the textbook . . . and applied on the ground."

When Kotok testified to the Appropriations Committee the following year, he presented a thirty-three page plan entitled *The Forest Research Program*. The plan was published in full in the committee hearings; typical of the process, Kotok offered to summarize the document for the members of congress. The first point was the effort to clarify the federal responsibility for forestry research and at the same time assess the size of program needed to "redeem" this responsibility. Given the long-running turf battles with the Department of the Interior, it is intriguing to note that he used the total federal forest ownership, not just the national forests, in his equation to justify the Forest Service research budget. He also pointed out that his research program served private ownership and that many forest problems crossed state lines, further justifying a viable federal presence.

In his testimony, Kotok acknowledged that Congress had asked the agency to consider consolidation of its research program. By his rendition, consolidation had begun in 1928 when the McSweeney-McNary Act authorized regional experiment stations. Hundreds of scattered research projects were then consolidated regionally. To Kotok the most "outstanding" consolidation had taken place since the end of WWII, when "programs at 103 experimental forests, ranges, and watersheds, plus several hundred supplementary studies on other lands were consolidated in the programs of only 62 research centers administered by the 11 regional stations." He did not see how further consolidation could be managed without reducing the agency's ability to conduct essential research on a nationwide basis.

Before turning to future needs, Kotok reported that cooperative research projects had increased from 140 costing $3.3 million in 1940 to 644 at $10.5 million a decade later. Cooperative efforts, too, had been authorized by the McSweeney-McNary Act and included agricultural experiment stations, state forestry agencies, universities, and private companies. He also pointed to the increased number of experiment station publications that helped get results used on the ground. Questions from committee members were revealing; were the reports "too academic" for field application, or did the agency's research involve itself in the controversy over federal regulation of private forest lands? Kotok's answers seem rather tangential to the questions, but the congressmen did not follow up. At times, the assistant chief seemed not to be attuned to the congressional culture, and he is not remembered as an effective leader.

As to future needs, Kotok's requests appear rather modest. Mainly, he sought more funding for fire research, especially for development of an improved fire danger meter. He also stressed the importance of resurveying the 423 million acres of forest lands inventoried earlier. The new data would establish insights on trends and would be much more useful than the initial measurement by itself. His questioners bore down on the benefits to the private sector of such information, and Kotok offered examples of how knowledge of wood supply would benefit a mill owner. This was the research chief's last testimony; he retired in 1951 and was succeeded by Verne L. Harper.

PART *3* RESEARCH EXPANSION BEGINS

Verne L. Harper Meets the Congress

Harper began his Forest Service career in 1927 with gum naval stores research in northern Florida. Just prior to being named assistant chief for research, Harper served five years as director of the Northeastern Forest Experiment Station in Philadelphia. Observers place him at least second to Earle Clapp for shaping Forest Service research. His fifteen-year tenure beginning in 1951, too, places him second only to Clapp for longevity.

Jamie Lloyd Whitten was chairman of the House agricultural appropriations subcommittee, and not only was he critical of agricultural research in general, he had lacked confidence in Kotok's research leadership in the Forest Service. Harper remembered how Whitten would ask Kotok "real tough" questions that even career researchers like Harper would have found difficult to field. When Kotok's answers were unsatisfactory, even glib some thought, Whitten declared that there was little justification to increase funding for a program hampered by weak leadership.

During Kotok's administration, pressure for additional funds for research centers by local and state witnesses appearing before House and Senate appro-

Verne L. Harper is pictured third from the left. Harper worked on naval stores research with the Starke Branch staff of the Southern Forest Experiment Station. (U.S. Forest Service photo.)

priations subcommittees had greatly irritated their chairman—Richard Russell in the Senate and Whitten in the House. In 1949, Russell had asked for a report on the extent that states cooperated with the Forest Service in needed research. Whitten found the Senate report to be unsatisfactory and in 1950 asked for a report on cooperative programs with the private sector, as well as the states. Harper, while director of the Northeastern Station, was assigned by Kotok to prepare the report to the Senate. Kotok's retirement made Harper responsible for the Whitten report. He told his associates that they would have to do their very best, and they came up with a "scheme" that would become known as "coop-aid research."

Not only was Harper new to the job and chairmen of key subcommittees hostile toward Forest Service research, but the Department of Agriculture's finance director watched nervously over his shoulder as he prepared the report, asking for "extra care and thoroughness." He expressed disappointment with Harper's effort, predicting that Whitten would also be disappointed. Harper stood firm and, with Chief Watts' backing, presented the plan at the next scheduled hearing. As it turned out, the chairman liked Harper's presentation and the proposed legislation that accompanied it. Although the "Whitten Bill" did not become law until 1956, from 1952 onward, his committee approved line-item requests for cooperative research. Some of the first projects looked at fire control issues in the Pacific Northwest in conjunction with universities, state agencies, and industrial protective associations.

Harper had fences to mend on the Senate side, as well. Appropriations subcommittee chairman Russell "displayed thinly veiled scorn and obvious sarcasm," as Harper asked for a small increase for fire control research. The senator insisted that a pine top worked just fine for beating out fires and the technology would benefit little from additional research. Russell also questioned why the Forest Products Laboratory developed so few new things with its large budget. He said that the same product samples had been shown to his committee for so many years that they "had been worn thin from repeated slidings across the witness table." Harper's acumen would eventually win Russell over, and the senator became a fire research advocate. Thus was the mood of Congress toward Forest Service research when Harper took over.

Relations with the USDA

There was the Department of Agriculture itself, where Forest Service research was not well known. To improve his visibility, Harper became active in the Agricultural Research Council, chaired by Byron T. Shaw, coordinator of department-wide research. Under the council, Shaw established in 1956 the Committee on Research Evaluation (CORE). This committee, whose members were to be drawn from each agency in the department engaged in research, were to spend fulltime for two months weeding out projects no longer needed and adding new research that was needed. Aside from this task, CORE members discussed new ideas about research, such as the man-in-the-job concept and pioneering units. Through the years, Harper borrowed CORE ideas for Forest Service research.

Harper also volunteered for other assignments for the department's research council. He soon found himself working as liaison to the National Academy of Sciences Research Council, further expanding awareness of Forest Service research.

Although Congress would become friendly, the White House and Bureau of the Budget saw agricultural research as descriptive and pedestrian, and perhaps not all that successful. Harper avoided debating agricultural research generally, which he thought had been enormously successful, keeping his focus on the Forest Service program. In fact, he agreed that the lion's share of forestry research to date had been descriptive and pedestrian, and rightly so. American forest managers in all sectors had needed applicable facts quickly; long term basic research studies would have to wait.

Harper reckoned that the time was right to increase emphasis on the more fundamental problems, and it was to be his term as head of Forest Service research that his successors would uniformly remember as a taking-off point for increased budgets and an ambitious laboratory building program. In his words, "we were ill-equipped [in 1951] with personnel, laboratories, and related scientific facilities to do as much fundamental research as we should be doing at the stations."

Harper thought that too much time could be spent evaluating research in terms of basic or applied. The measuring systems were too subjective, and what was basic research in one generation could be judged later as inadequate. For him, good examples of earlier, fundamental research were Gisborne's work on fire, Leo Isaac's Douglas-fir studies, Lloyd Austin's investigations in forest genetics, grazing studies by A.W. Sampson, and naval stores and watershed research in the South.

Robert Buckman, later deputy chief for research, remembered vividly the "generational switch" during the 1950s. As a freshly minted Ph.D. from the University of Michigan, he was well-steeped in statistical design and analysis, while scientists much his senior felt that descriptive studies were still adequate. Following a bit of tug-of-war, Buckman received a green light to embark on a major investigation that used statistical design as its centerpiece. Thus, research standards laid out in the 1930s were coming into wider usage. He is a good representative of the new generation of scientists; more and more new hires would have a Ph.D. in an appropriate field of science, and at the same time the agency encouraged its other researchers to return to graduate school for the advanced training that was becoming essential. By the 1960s, the Ph.D. degree would be commonplace throughout the Forest Service research establishment.

Some researchers who lacked advanced degrees were concerned by the emphasis on obtaining additional training. Harper remembered being "threatened once by a small group of young researchers with bachelor's degrees in forestry who bluntly said they would leave the research organization and seek employment elsewhere rather than be forced to acquire higher degrees." He responded by encouraging them to leave "if they did not wish to become qualified to do sophisticated research." By 1960, 13 percent of the research work force that lacked the Ph.D. degree were enrolled in university training, assisted by flexible work schedules and the Government Employees Training Act of 1958. This program allowed about 10 percent of the researchers each year to gain advanced training; it would end in the late 1960s when no longer needed.

The general climate was supportive of research, with a Sputnik-driven need to "keep ahead of the Russians" loosening congressional purse strings. By the 1970s, thirty-five fields of science were represented in the Forest Service research organization; about half of the new hires were trained in forestry schools and the other half from science and social science departments of the university.

Cultivating Congress

Harper was tireless when cultivating congressional relationships; that was the only way to get increased appropriations. "All told, I believe I saw at least one senator from each state and two or more congressmen from a region of each of our stations. In most cases I was cordially received and in some cases I was asked many questions about our research needs." As it would turn out, one of his most important contacts was Senator John Stennis from Mississippi.

Harper was still new to his job when Stennis sent a request to the Forest Service for information about bristlecone pine. Harper used the opportunity to meet Stennis and responded to the request in person. The senator had just returned from California, where he had visited the Institute of Forest Genetics. He had brought back a specimen of bristlecone pine and wanted to know more about it. He thought it would be a good thing to use in the Sunday School class he taught; here was a piece of wood that had been alive during biblical times. It was a good meeting, and Harper learned that Stennis owned a small tree farm in Mississippi and was familiar with forestry in general. He was a staunch supporter of research and also was on the Senate appropriations committee. Two other senators who were forestry research supporters were Carl Hayden of Arizona and Robert Byrd of West Virginia. The three would be consistently instrumental in awarding Harper's ever-larger appropriation. In fact, research would routinely receive from Congress more than requested in the President's budget.

By the time Harper retired in 1966, there were Forest Service research laboratories in forty-five states. Each unit had required congressional involvement for construction and again for staffing. This high amount of activity assured that both houses of Congress were familiar with research and could be counted on for support at least when it involved a particular state or district. Harper became so successful with Congress that he began to suspect the Bureau of Budget reduced his appropriation request even more, in anticipation that Congress would overrule it anyway. But, there were some who felt that the deputy chief had been too aggressive, and those who followed Harper found internal fences that needed mending.

A Variety of Programs

Larger budgets, new facilities, and increased staffing meant an expanded research effort. Harper oversaw the addition of formal programs in pathology, entomology, engineering, marketing, habitat, and recreation research, a doubling of fields of research. Of course, work had gone on in these "new" fields, but not on an organized basis. For example, wildlife management research in the Forest Service began in 1946, although authorized in 1928 via McSweeney-McNary, and recreation had been studied as part of work in economics and wildlife.

Entomology and pathology came through transfers from other agencies in the department, but it looked as though range would go the other way into the Agricultural Research Service. Range was the maverick that had remained outside of Earle Clapp's reach until 1926; in 1953 it apparently was to be lost to the Pasture Branch of the Crops Research Division of ARS. Harper huddled with his senior staff to develop a strategy to retain as much of range research as possible. They learned that the transfer had been proposed by the livestock industry to a sympathetic Ezra Benson, the new secretary of agriculture. Since there seemed little scientific logic behind the transfer, Harper pressed ahead. He suggested that only portions of range research that dealt with treeless areas, such as the Jornada Experimental Range in New Mexico that the Forest Service had acquired in 1915 from the Bureau of Plant Industry, be transferred. To be retained in the Forest Service would be research on forest and related ranges, reseeding or revegetation of range for wildlife habitat, range ecology, and plant control by grazing management and fire. The secretary's office accepted the recommendation, and thus little range research was transferred to ARS.

Early in his term of assistant chief for research, Harper saw the Research Branch being challenged in its overseas role. Chief and staff heard a proposal to move international forestry out of research, for almost all of the activities were related to technical assistance. But, Chief Watts rejected the proposal feeling strongly that Harper's branch was the proper place. Showing a degree of provincialism, the effort was labelled the Foreign Forestry Unit until 1961, when Robert K. Winters became the director of Foreign Forestry Services. The name was changed to International Forestry in 1965. The efforts increased, remaining largely in the Branch of Research until 1991 when International Forestry received its own deputy chief.

Recreation research was something that Harper had advocated even while he was an experiment station director. He remembered, however, that he "could not arouse the interest of anyone—industrialists, conservation organizations, watershed councils, or others—to publicly support this kind of research." When he got his first real recreation research money in 1955, he promptly contacted Samuel T. Dana, former dean of the Michigan School of Natural Resources and much earlier director of the Northeastern Experiment Station.

Harper commissioned Dana to make a problem analysis for forest recreation, the first step in launching a formal program. Dana's persuasive powers were sorely tested, as he encountered opposition to recreation research. He prevailed, however, and the 1960 budget carried recreation as a line item within forest and range management. Harper believed that the way had been cleared a bit by the 1958 creation of the Outdoor Recreation Resources Review Commission and strong support from senators Hayden and Stennis for the entire research budget. In the 1961 Forest Service annual report, recreation led off the research section, and the reader learned of studies on campsite spacing, site protection, facilities design, and wilderness use.

G. A. Pearson had been named director of the Fort Valley Experiment Station in 1908, since renamed as an experimental forest. In 1935, Pearson asked to be relieved of administrative duties so that he could devote full time to research. Ponderosa pine and its various properties was his main interest; in 1950 his Management of Ponderosa Pine in the Southwest would appear as a 218-page publication, the capstone of his scientific career. As had Thomas Croker with longleaf pine and Issac with Douglas-fir, Pearson and others like him produced monographic literature, species by species, adding to a basic fund of knowledge that would be used universally.

Trees were of interest in the Great Plains, too. In fact, in 1873 Congress had passed the ill-fated Timber Culture Act as a variation to the 1862 Homestead Act. Under the latter law aimed at settling the West, one could gain ownership of a 160-acre plot of public prairie by planting trees on a portion of it. According to the science of the time, trees affected climate, as evidenced by the observation that wherever there were trees there was ample precipitation. Not only would rain come to the dry prairies, but the trees would also provide firewood, fence posts, and lumber.

There had been more recent interest in the prairies during the 1930s with implementation of the ambitious shelterbelt project that put people to work during the depression planting trees in rows to shelter farmhouses and crops. By 1953 it was time to examine the shelterbelts, and Forest Service researcher Ralph Read found himself in Lincoln, Nebraska; his office was "a table and chair placed in the hallway just outside the office of the chairman of the

Samuel Trask Dana began working for the Forest Service in 1907, serving as a director of the Northeastern Experiment Station from 1923–27. He left the agency to become Dean of the School of Forestry and Conservation at the University of Michigan, a post he held for 24 years. (Forest History Society photo.)

Horticulture Department in the Plant Industry Building." In two weeks he was able to share a real office. His first task was to prepare a problem analysis. Three years later he was joined by a pathologist, silviculturist, and soils specialist, as he worked on applied genetics studies.

There was also interest in Alaska's forests. In 1927 Thornton Munger had travelled north from the Portland experiment station for a six-week tour of Southeastern Alaska to explore forest conditions. The following year Raymond Taylor headed a two-man research program out of Juneau. In 1948 this modest effort was expanded into the Alaska Forest Research Center; one of the first studies was in the Maybeso Experimental Forest on the impact of logging on salmon spawning grounds. But the big break came in 1954 when Congress amended the McSweeney-McNary Act to fund forest research in territories, in addition to the states. By 1961 the Northern Forest Experiment Station was created. Six years later it merged with the Pacific Northwest station and was renamed the Institute of Northern Forestry.

In 1957 Carl Ostrom moved from the Asheville experimental station to

Ralph Read explains how white oak growth on the Slyamore Forest quadrupled due to timber stand improvement work. His audience includes author and lecturer Harold Sherman; the former director of the Arkansas Resources and Development Commission, Hendrix Lackey; and U.S. Congressman Wilbur Mills. (U.S. Forest Service photo).

Washington, D.C. to become the director of forest management research. To acquaint himself with the national program, he visited all of the stations. In California the number one problem was how to improve techniques to secure natural and artificial regeneration. In fact, at each station Ostrom noted that forest regeneration studies ranked high, whether it be genetics investigations at Rhinelander, Wisconsin, or seed-soil studies in Portland, Oregon. The forestry novice might well have assumed that by mid-century foresters would have learned all they needed to know about planting trees. But that was not the case, and scientists worked to improve the process.

A natural outgrowth of the requirement for statistical design was the addition of statisticians to station staffs. Floyd Johnson was the first, beginning in Portland in 1946, followed the following year by C. Allen Bickford at the Northeastern station, and G. Luther Schnur at the Central States Station. By the 1950s, all stations sought statistical specialists—soon to be called biometricians—and at the end of the decade the roster was complete.

Timber Supply

Timber's prominence in the early 1950s was not surprising; for a half century the conservation movement had been primarily driven by concern that not only supply was diminishing but that too much was wasted in the course of utilization. In 1952 the Forest Service undertook what it called a Timber Resource Review, which would include all holdings. The project was assigned to assistant chief and economist Edward C. Crafts, who drew heavily upon expertise in the Research Branch.

During the review, the industry turned to its friends in the Eisenhower administration to limit the study to timber supply and not include broader issues of policy. When the agency proposed to establish social, political, and economic forms of reference, the industry's primary trade association labelled the effort "pure and simple hogwash." The industry weighed in by retaining John Zivnuska, an eminent forest economist on leave from the University of California, to analyze a 1955 preliminary report. Zivnuska expressed a variety of concerns, especially about the Forest Service effort to predict conditions forty years distant. The agency responded by assigning James Rettie from the Division of Forest Economics Research to analyze Zivnuska's evaluation. Rettie stoutly supported the preliminary data and insisted that predictions of supply and demand were integral to the study. The public and private debate was often bitter, and the end result was a limiting of the investigation to timber supply. Research was still on the front line.

Chief Forester Richard E. McArdle presents Secretary of Agriculture Benson a copy of Timber Resources for America's Future. *This publication repre-sented the most complete study of the nation's timber resources. The interested onlookers, whose futures are intimately tied up with the future of their country's forests, are Cub Scout Steven Lewis; Brownie Marguerite Harris and 4-H girls Linda Gray and Jean Gooding. (U.S. Forest Service photo.)*

The *Timber Resources Review* was published in 1958 to generally wide ac-ceptance. It included by far the most comprehensive and sophisticated de-scriptions of supply, with predictions to the year 2000. Significantly, included was a statement by Chief McArdle that "there is no 'timber famine' in the offing although shortages of varying kinds and degrees may be expected." With Pinchot's timber famine finally put to rest, foresters could look to other issues.

Forest Research Advisory Committee

To gain a broader acceptance of its research program and to benefit from di-verse perspectives, in 1952 the secretary of agriculture established the Forest Research Advisory Committee. The committee, nominated by the Forest Service, included academics and state representatives, as well as those from the forest industries. The group met and toured research facilities where they could see firsthand studies of mutual interest.

We have seen that the forest industry was leery of the Timber Resources Review, and it was no less leery about the advisory committee. An Industrial reviewer branded the committee's 1953 statement of goals as "just another parade of propaganda to justify further expansion of the USFS." Others recanted depression-era support for the Forest Products Laboratory by stating opposition to applied research that was better and more economically accomplished by the industry itself. Marketing research was to be avoided, too, as "these fellows are enough of a problem in the forests—let's keep them out of the market place." Recreation research was seen as better conducted by the states.

Despite this sour startup, the committee continued its work. For example, during four days in October 1958, the group observed conditions and needs as it travelled from Asheville, North Carolina, through Georgia and on into Florida. The entire November 1958 issue of *Forest Farmer*, the committee's dinner host in Athens, Georgia, was devoted to forestry research in the South.

The issue led off with Harper's foreword where he estimated that total research expenditures for forestry were three-tenths of a percent of consumer sales, while the figure for all of agriculture was five tenths. The average for all industry was 1.3 percent of consumer sales. Thus, funding for forestry research of all kinds would need a four-fold increase to match the average, an ambitious goal indeed.

Philip Briegleb and Joseph Pechanec, directors of the Southern and Southeastern stations, followed with a status report on southern forestry research. There had been a breakthrough with the development of effective bird and rodent repellents for pine seeds, needed to efficiently produce more than a billion seedlings planted each year. At the other end of the rotation, cooperative studies by the Forest Service and industry had produced much more efficient processes to convert sawmill waste to pulp chips, saving $24 million in value from what previously had to be burned up or carted off. Too, there had been the 1954 opening of the Southern Institute of Forest Genetics and the 1957 startup of the Southern Forest Fire Research Laboratory, the latter constructed by the State of Georgia and given to the Forest Service for operation.

Cooperative research was the name of the game; more than a third of the Southern and Southeastern experiment station budgets came from state and private sources. A good example was hardwood research. Hardwoods are abundant in the South and much of it is of low commercial quality. In 1953 forty industrialists and hardwood landowners formed the Southern Hardwood Forest Research Group. Three companies—Anderson-Tully, Chicago Mill and Lumber, and Crossett—set aside land for experimental forests.

Much of the money raised by the group supported the Southern Station's Delta Research Center at Stoneville, Mississippi. Cottonwood cultivation, tests of even-aged versus all-aged management in mixed hardwoods, and thinnings in sweetgum, and black willow showed promise for commercial application. There was also much interest in the soon-to-be-published USDA southern hardwoods handbook that would support foresters responsible for both management and inventory.

From time to time the Forest Research Advisory Committee filed reports with the secretary of agriculture, and participants felt that the enterprise was worthwhile. However, times and needs change, and in 1970 the secretary discontinued the committee because it became seen as an agent to lobby for increased appropriations.

Fire and Water Research

Fire researchers increased their efforts to understand, prevent, and suppress wildfires. Keith Arnold, who would eventually rise to deputy chief for research, was asked to lead Operation Firestop. Civil defense, Department of Defense, state agencies, and municipalities joined with the Forest Service to learn more about the air-dropping of chemical retardants and water and the use of helicopters. During the same period in the Rocky Mountains, Operation Skyfire undertook to learn how to prevent the lightning strikes which caused 90 percent of that region's forest fires. The Forest Service would have fire research laboratories in Macon, Georgia; Missoula, Montana; and Riverside, California.

Water research focussed on the watershed, sometimes forested but sometimes not, at least not commercially. The Rocky Mountain Forest and Range Experiment Station studied both kinds. The Fraser Experimental Forest had been established in 1937, and most of its water was diverted to Denver. The station gave high priority to capturing more water from snowmelt. Undoubtedly the most famous watershed photograph ever taken was of one such study, where Fraser Forest's Fool Creek watershed was clearcut in a variety of patterns, with careful monitoring of runoff quantity and quality. The success of these studies sparked interest in Arizona.

There was opposition to the direct application of treatments developed high in the Colorado Rockies to a rather more arid section of Arizona, even though advocates insisted that it would work. Senator Hayden, chairman of the appropriations committee, asked Chief McArdle if tests would be possible on a selected watershed that represented Arizona conditions. The chief's ready "yes" brought an increased appropriation for the Rocky Mountain Station; water researchers picked the Beaver Creek watershed on the Coconino National Forest. As on the Fraser, treatment consisted of removing vegetation in a variety of patterns and monitoring the effects. The results were an ability to increase water yield in an area where water was always scarce.

In Utah scientists looked at the situation from a different angle. In 1912 adjacent drainages, efficiently named Watershed A and Watershed B, were carefully examined by Harold Croft to determine baseline conditions. Although observations were continued for many decades, the first study was related to flood attenuation. By 1920, using controlled grazing to reduce ground cover, investigators reported conclusive evidence that summer floods would result from overgrazing. In fact, Clarence Forsling was one of the several subsequent users of the adjacent watersheds who were able to build on Croft's initial project. In Colorado and Arizona, the aim was to increase the amount, while in Utah it was to affect distribution.

Often it is possible to generalize from research results—indeed it is usually a scientific goal—but the diversity of conditions even more often made it necessary to study topics regionally. What applied in one situation would not necessarily apply elsewhere. For such reasons, the humid South, too, needed to study water conditions, as information from the arid West appeared rather exotic to observers east of the Mississippi River. Thus, in 1933 the Asheville station created the Coweeta Experimental Forest to capture the full benefit of hydrologist Charles R. Hursh who had been with the station since 1926. He would retire from Coweeta in 1953, but like drainages A and B in distant Utah, the North Carolina watershed research area offers still-used baseline data and facilities.

Forest Products Laboratory

The Madison laboratory had served the country well during the two world wars, and its ties to the military continued through the Cold War. In 1954 Harper testified to Congress that only 160 of the 350 Lab employees were supported by the regular Forest Service budget; the balance were covered by defense-related contracts. An example was the newly developed "sandwich" material—thin strong skin over lightweight core—that was vital for the manufacture of jet fighters and missiles. The money was good, but it also meant rather too much time was spent on applied research at the expense of much needed basic. By the end of the decade the Forest Service allocation to the Lab more than doubled, the defense portion was reduced to 20 percent of the total, and there was a concurrent shift of attention to basic research.

During this same period there were, of course, important civilian advances. There was the cold-soda process for hardwood pulp that by 1957 four companies were using to produce two hundred tons daily. The hardwood pulp was manufactured into corrugated board and printing paper. Too, there was direct involvement with southern manufacturers to develop plywood from pine. Also developed at the Lab was a low-cost veneer flooring and FPL Natural Finish for redwood and cedar siding. Research continued on glue-laminated construction and soil treatments for termite control. Yet another example was two companies using a Lab process for Impreg, a key to making stable tool die models for the auto industry. One of the auto companies calculated that it was able to shave $2 million per year off its retooling cost.

Administrative Advances

While scientists tackled directly the many pressing issues at hand, research administrators looked more broadly at the situation, including the investigators themselves. Assistant Chief Harper sought ways attract and keep top scientists. Through his participation in CORE meetings, he learned of the man-in-the-job concept, which allowed researchers to be promoted on the basis of their contributions to knowledge rather than by accepting administrative responsibilities. The Civil Service Commission gave approval in 1957 for a pilot program for agricultural research. It quickly proved successful, but it would not be until 1961 that the Forest Service personnel office authorized

Inspecting paint test panels of the Forest Products Laboratory. The panels are located on the University of Wisconsin farm at Madison, Wisconsin. (U.S. Forest Service photo.)

its application. From that point on, the accomplished scientist would not have to trade the bench for the desk in order to advance.

As we have seen, from the CORE meetings Harper got the idea for pioneer units that could be made available to the handful of truly outstanding scientists. In these cases, a whole program could support the efforts of a single individual, who would work essentially without supervision. The first pioneer unit was established in 1960 at Berkeley around the mensurational genius of Louis R. Grosenbaugh. Other examples were physiologist Philip Larsen in the Lake States and utilizationist Peter Koch in the South. If an especially promising new recruit would question future opportunities, a supervisor would only have to point to the pioneer units. For the very best, the sky was the limit in Forest Service research.

By the mid-1950s, some began to question the concept of the research center, which were established at some distance from the experiment stations themselves. Deficiencies included superficial research, some center leaders were becoming too autonomous, and certain overhead tasks were duplicated at the station. The agency asked the consulting firm of McKinsey and Company to examine its entire administrative structure. Among its many recommendations was the continuation of the research center concept, which Harper found unacceptable on the grounds that it would perpetuate the bad situation of depending upon the lowest paid employees for doing research under the center administrator. In a major decision, chief and staff rejected that recommendation and adopted instead a system based upon projects. A project-based organization allowed transfer of many administrative tasks to the station and increased the depth of research. Too, geographic boundaries became less significant, while the subject at hand became more significant.

Another change dealt with the myriad publications that the stations and Lab published each year. Research findings are not put to use if the reports are not read, and there was an agency-wide effort to make the publications physically more attractive. Also, editorial specialists, such as Robert Wray at the Central States station, E. W. Shaw at the Rocky Mountain station, and Edwin V. H. Larsen at the Northeastern were signed on. Eventually all stations had an editor. Not only did station editors critique manuscripts, but they also conducted training programs for the scientist.

Forestry Research Plan

There were many reports, but in 1961—this one on research alone—a Forest Service report looked ahead ten years. Three years earlier the agency had issued a Program for the National Forests, which included a short section on research. With associate deputy George Jemison playing an instrumental role, the later report was designed to showcase research and make a pitch for an orderly three-fold expansion over the coming decade, plus a hefty building program to house the increased projects and personnel. Although used internally, the report was bottled up in review for three years, receiving final approval in 1964.

The report began with an overview of how much previous research had already been incorporated into daily forest management activities. Improved timber harvest techniques, reforestation, range quality, fire suppression, and treatments for insect and disease infestations were just some of the management advances that could be directly attributed to the agency's research program. There had been substantial progress made on the products side, as well; improved utilization in the mill, low-grade hardwood usage for pulp, and enhanced byproducts usage in the paper industry—all led to a stronger forest economy. All this advance was well and good, but increased demands on the nation's forests and increased public concern about management quality meant that much more knowledge was needed if the forests of the future were to be healthy and able to supply a projected two-fold increase in demand for forest resources by the year 2000. Two-thirds of research was being conducted by the forest products industry with the bulk of the balance by the Forest Service. Very much needed was a major increase in research by universities.

The report had quickly cleared the Forest Research Advisory Committee, but the White House science staff wanted additional review by the committee of scientists that it had required to be established by the Department of Agriculture. Harper remembers that "The White House science staff took a dim view of the scientific caliber of the USDA advisory committees . . . and it also had a low opinion of agricultural research generally. . . ." Too, the report was to receive careful scrutiny from the Advisory Committee on Agricultural Sciences appointed to advise the secretary. This committee clashed with Harper over predictions of inflation and the long time-frame. Harper held firm on his plan, and eventually it cleared the final hurdles. Senator Stennis, strongly supporting the proposed research facilities, was instrumental in keeping implementation reasonably on course each year in the Senate.

McIntire-Stennis Law

The Department of Agriculture had established the Committee on Research Evaluation (CORE) in 1956. Its purpose was to identify research areas for either curtailment or expansion and also new areas for emerging problems. CORE'S draft report was circulated to state agricultural experiment stations and forestry schools; the final report was published in 1960. Harper observed during the review sessions that forestry deans were included for the first time in research planning discussions between the department and land grant schools. For obvious reasons, such discussions had focussed on soybeans, peanuts, cattle, and so forth, but now forestry was included. Once again he saw an opportunity to strengthen cooperative research. The Whitten Act of 1956 had been a good start, but more needed doing along these lines.

In early 1962 Harper was visited by the Legislative Committee of the Organization of Agricultural Experiment Stations. The committee wanted Harper's advice on how best to seek enabling legislation that would give the state experiment stations much needed funds for forestry research. The committee suggested that either the earmarking of Hatch Act funds or new legislation would be acceptable. Harper quickly opted for new enabling legislation.

With the backing of the forestry schools and state agricultural experiment stations, Harper canvassed chief and staff, and received their support as well. Further, Edward C. Crafts who normally handled the agency's legislative assignments, suggested that Harper could handle this particular bill on his own. Harper took the lead to steer the bill through the Bureau of Budget, the White House, and congressional committees. One tactical question answered itself when Congressman Clifford McIntire from Maine requested a forestry cooperative research bill. The request was not that much of a surprise, because the University of Maine's forestry dean, Albert Nutting, had been especially active during the earlier review process. Also, Harper had told anyone on the Hill who would listen that such a bill was available in draft form.

McIntire approved the bill with a few, minor changes, which Harper quickly reported to Senator Stennis, who had cosponsored a similar bill. Stennis had also invited senators James Eastman and George Aiken to join him, and they did. Thus the bill seemed to be in good shape in both houses at this early stage.

The first snag came from the forest industries. Still bruised from the Forest Service's several-decade campaign for federal regulation of the private sector, the industry was leery of having the Forest Service directly linked to financial support of state forestry schools. So the issue quickly became who would administer the funds. Harper suggested that the Cooperative State Research Service, a USDA agency, could be administrator and would meet industry's objections. To avoid lengthy delays that amending the bill might entail, he further suggested that the legislative history could state that CSRS would be administrator, and this strategy too received general approval.

As the date of the scheduled hearings approached, McIntire asked Harper to find out why the Budget Bureau had yet to release its report. Looking into it, he found that it was bottled up in the White House and not the bureau. Harper asked for and quickly received an appointment to meet with the White House staffer who had declined to release the report. Accompanied by two USDA associates Harper met with the Budget Bureau staffer and the person from the White House. His strong objection was that research funds ought not be distributed across the states; the superior institutions should receive most of the money, because that approach would yield the best results. He also did not think much of agricultural research in general, and to him, forestry was a kind of agricultural research.

The discussion seemed to reach an impasse, and Harper had just about given up, when the opposition collapsed. The White House would stand aside; the report could go forward. Subsequent testimony went as planned, there was no opposition and the legislative history clearly stated that the secretary of agriculture intended to use CSRS as administrator. One final hitch was whether forestry programs at private universities, such as Duke and Yale, ought to be included. Although the proposal had merit, an amendment of this nature would have delayed passage, and the proposal was turned aside. The McIntire-Stennis Act became law on October 10, 1962. Harper had done a lot, but he well knew that it had happened because of strong support and advocacy from state agricultural experiment stations, forestry schools at land grant colleges, and friends of forestry research in Congress.

McIntire-Stennis opened a broader avenue for the Forest Service to cooperate financially with the nation's forestry schools that were located at land grant universities. As Harper had pointed out privately to Stennis, forestry graduate students and faculty had needed skills that could be brought to bear on key problems, often less expensively than the agency itself could bring to the task. Further, training graduate students in forest science would assure a supply of scientists that the Forest Service would need in the future. If 1952 can be viewed as the take-off point for the growth of Forest Service research, 1962 has similar relevance for university-based forestry research. By the end of the decade, about sixty universities were receiving McIntire-Stennis funds, and the allocation had more than tripled to $3.5 million from an initial $1 million. By the 1990s the amount would reach about $17 million.

PART  **4** **RESEARCH AND THE ENVIRONMENT**

Silent Spring in a Noisy Decade

Rachel Carson's *Silent Spring*, published to wide acclaim in 1962, provides a convenient starting point for the environmental movement. The key thrust of the movement was to go beyond the conservationist goal of prudent husbanding of natural resources to include quality of life and to place value on things that the marketplace had yet to recognize. It was a complex time, given the parallel emphasis on civil rights and wilderness preservation, and Forest Service research reflected the times.

Scientists who read *Silent Spring*, whether they approved of it or not, generally agreed that Carson had revealed little that was new; the revelations were to the broader public. The persistency of DDT had been well documented in the literature since the late 1940s, and studies on its effects had repeatedly received congressional support. Following *Silent Spring*, Congress allocated the largest single appropriation for a specific program to date, $29 million to intensify the investigations on biological controls. Forest Service research also received a big increase, and shortly entomologists reported the successful biological control of the roundheaded pine beetle in New Mexico and the tussock moth in Oregon. The larch casebearer and dwarf mistletoe would also meet their biological match, and on down the list of insects and disease that plague forests. Scientists were also seeking a replacement for DDT for those situations that could not yet be treated biologically.

Minorities in the Workplace

Even more vividly than the impact of *Silent Spring*, researchers remember the intense pressure to hire blacks throughout the government. George Jemison had been associate deputy chief under Harper and moved to the top research spot when Harper retired in 1965. The department had issued a quota to the Forest Service, and it fell to Jemison and the research branch to be the most aggressive recruiter of blacks, especially at the higher levels. Basically there was a freeze at the GS-14 level and above, and black candidates were to be given special consideration. Jemison also visited black colleges to spread the

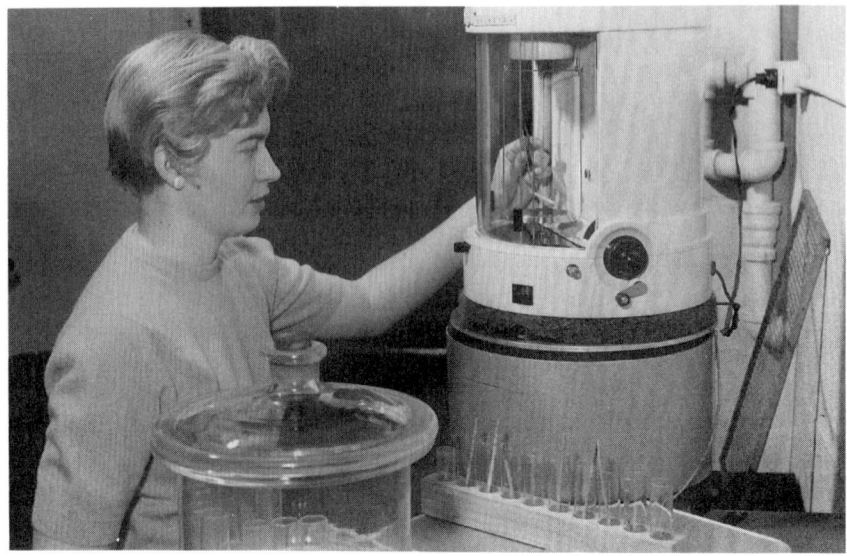

This research worker is using the precise measurements of borings to determine the relationships between factors such as growth rate, specific gravity, and the percentages of springwood and summerwood. (U.S. Forest Service photo; Forest Products Laboratory.)

news about research opportunities in the Forest Service, and he established a research position at Alabama's Tuskegee Institute. He also constantly monitored, experiment station by experiment station, progress in minority hiring in the field. Each of his successors would tell a similar story; the effort to achieve the appropriate level of what is now labelled cultural diversity was to receive high priority, as the essentially all-male organization began to bring in women professionals and minorities in ever-increasing numbers.

Research at Mid Decade

Recreation researchers studied wilderness use in the Boundary Waters Canoe Area in Minnesota and on the Three Sisters Wilderness Area of Oregon. The Wilderness Bill had been hotly debated for eight years before its passage in 1964; major issues were what criteria were necessary for satisfactory wilderness experience, who used wilderness, and was wilderness a "resource" as stated in the act. Researchers found in Oregon that 91 percent of the wilderness visitors were Oregonians—users were "local." In Minnesota it was found that Boundary Waters' use was considerably above previous estimates and that visitors tended to concentrate in certain places—users might want a wilderness experience but did not want to do it alone. Meanwhile, in Utah researchers were investigating the possibility of "micro-wilderness" areas to determine whether visitors would be satisfied by still rustic but less isolated experiences. In this latter case, there would be potential of many more wilderness setasides more convenient to human populations, if not limited to relatively pristine areas of a half-million acres or more.

Fire researchers continued to study the effects of fuel moisture, topography, and weather conditions on fire behavior. This work was greatly bolstered by the three fire laboratories that allowed careful replication of treatments in a way that field studies had not. Project Skyfire continued the practical goal of enabling land managers to prevent lightning strikes. Airborne infrared scanners, developed by the military and tested by fire researchers, could enable observers to "see" a fire though heavy smoke. Following several years of study via Project Firescan, scanners were in common use by fire control agencies. Too, fire-retarding chemicals and compounds, developed and tested, had become a vital suppression weapon. Two major contributions were the growing ability to model and predict fire behavior and the development of a scientific base that would allow elimination of the 10 a.m. fire policy for federal agencies. In the fire-driven world of American forest management, this latter policy shift was of great significance.

Under the general field of forest management research were included genetics studies to develop "improved" trees—longer fiber length, uniform wood density, and disease resistance. Silvicultural research yielded a new guide for managing hardwoods in the Central States, sought optimum stock

Controlled pollination of western white pine at the Institute of Forest Genetics in Placerville, California. The bag prevents random pollination; the syringe contains pollen from a selected male, to produce a superior tree. (U.S. Forest Service photo.)

ing levels for the South's loblolly pine, and measured nutrient loss brought about by timber harvest.

Economists noted that lumber was being displaced by plywood in home construction, a shift of significant consequence to those who grew trees and milled logs. Too, they predicted that by 1985 in the Pacific Northwest, shifts in technology would reduce the number of loggers by a quarter and mill workers by nearly half. Also in the Pacific Northwest, log exports, largely to Japan, were causing higher domestic prices and creating a hardship in some sectors of the forest products industry.

Environmental Forestry Research was a new heading in the 1967 chief's report. Included here was work on air pollution and its effect on tree mortality. In only a few more years, there would be much concern about cleaner air and related issues. This early research enabled the Forest Service to achieve a much broader base for its later investigations.

The peculiar organization of the federal bureaucracy in combination with the laws of Congress meant that the Forest Service in the Department of Agriculture studied wildlife habitat, but it was the Fish and Wildlife Service in Interior that was charged with studying the animals themselves. Scientists are used to collaborating across disciplinary and cultural boundaries, so that the separation of duties may be more apparent than real, but no doubt the situation caused some inefficiencies as the Forest Service focussed on wildlife in terms of food and shelter. Also, wildlife research was joined administratively with range research, stemming from the manager's traditional need to "balance" the competition for forage between wildlife and domestic livestock. Another peculiar grouping that made sense to the agency was administratively placing fire research—largely a physical science—with the biological sciences of insect and disease research. Through the years the combinations have been adjusted in accord with shifts of emphasis and changes in personnel.

International forestry, although under research, was largely an extension effort. The training of foreign scientists and cooperating with the forestry programs of FAO and USAID were key activities, as was the translation into English of thousands of pages of studies conducted in other countries, paid for quietly by the Central Intelligence Agency. *Terminologia Forestal*, a Spanish-English terminology guide, would boost international activities, as would the hundreds of foreign nationals who received technical training each year. Also, the deputy chief traditionally played a leadership role in the International Union of Forestry Research Organizations. An interesting and fruitful wrinkle was having access to the so-called PL-480 monies. Through this mechanism, the Forest Service could design and direct research in soft-currency countries that had received economic aid from the United States. Thus, there were U.S. directed studies in India, Italy, Finland, the Philippines, Israel, and Poland, as well as other nations. Essentially, the PL-480 studies constituted the research that took place in international forestry.

Administrative Issues

Jemison retired in 1969 after less than four full years as deputy chief for research. During his short tenure, he had been able to take advantage of the momentum that he and Harper had built. The research budget in 1960 had been a little more than $15 million and by 1965 when Jemison took over it had doubled to $31 million. By 1969 it was $40.7 million. Inflation lessens the effect of the larger numbers, of course, but nonetheless it was a time of heady growth.

The growth was especially noticeable in the field, with the construction of experiment station headquarters and laboratories. In 1959 the only Forest Service laboratory was in Madison. That year, Congress appropriated $2.5 million for construction of additional facilities. They added another million in 1961, $5.2 million in 1962, $2.6 million in 1963, and similar hefty amounts through the decade for a total of $28.6 million. Less and less often would researchers conduct their work in federal buildings such as courthouses and post offices, or share quarters with university programs. State of the art facilities provided workspace and supported the operation of the new technology that had become so essential to a wide range of scientific investigations. The number of scientists grew as well to nearly one thousand, a number that would be maintained until the 1980s.

To Jemison, research was not yet fully equal with the other Forest Service programs in National Forest Administration and State and Private Forestry. Station directors were in the federal salary grade of GS-14, while regional foresters were GS-15. When Jemison examined the skill levels needed and breadth of responsibility for directors, he believed that they were equivalent to their counterparts in the agency's administrative branches. He prepared detailed descriptions of duties and convinced the classification officer that it was "not only fair but essential if we were to keep any semblance of equal pay for equal responsibility." He was also persuasive at the next level up in the department, but it was a different story at the Civil Service Commission. Jemison was asked to place the directors in order of priority; presumably only some would receive the increase. He held firm, however, and all nine direc-

These are the first three directors of the Lake States Forest Experiment Station. Raphael Zon, the first director from 1923–1944, is pictured in the center. At the right is Elwood L. Demmon, who served from 1944–1951. Murlyn B. Dicker-man, at left, served the station from 1951–1964. (U.S. Forest Service photo, 1951.)

tors were moved up. As he remembers, "it was quite a chore," but well worth the effort. With inauguration of the Senior Executive Service in the early 1980s, the compensation structure for regional foresters and directors, and their deputies, would shift again.

The Environmental Decade

When George Jemison was director of the Pacific Southwest Experiment Station in Berkeley, Keith Arnold worked for him as director of fire research. Then Jemison was moved to Washington to be associate deputy to Harper; Arnold was named station director at Berkeley. Then Arnold, too, was brought to Washington to head the Division of Forest Protection Research, but from there he moved to Ann Arbor to be dean of the University of Michigan School of Natural Resources.

M. B. Dickerman had been Jemison's associate deputy, and most saw him as the next deputy. However, for personal reasons, Dickerman removed himself from consideration. Lacking an obvious in-house candidate, Chief Edward P. Cliff was concerned that the secretary would bring in someone from the outside for the top research spot. The chief called Arnold, who agreed to return to Washington, and in May 1969 he became deputy chief for research.

Arnold was bristling with ideas, and with the break in service caused by his stint at Ann Arbor, he had inherited no particular commitment to Harper's programs. He did believe, though, that he had benefited greatly from the momentum that was in place.

The next year there would be the National Environmental Policy Act, and President Nixon would use his authority to create the Environmental Protection Agency. As the president penned his signature to NEPA, he announced that the 1970s would be the environmental decade. Although the nation had responded to the environmental impulse in a variety of ways during the 1960s, now the government set out to put the federal environmental house in order. Forest Service research was responsive to the opportunity. Arnold's goal was "to move forestry research to the cutting edge of environmental policy."

At the Cutting Edge

Arnold aimed to increase emphasis on recreation research and to develop basic studies of ecosystems. The agency was already studying most of the American ecosystems and was the largest single employer of ecologists. He also wanted to attract young scientists with broad interests. At his first regional

R. Keith Arnold in Yugoslavia, 1970. (Forest History Society photo.)

foresters and station directors meeting as deputy chief, he "stressed the increasing concern about the environment and how it impacted forests." He felt that the chief and other deputies were in agreement that the Forest Service should move more quickly, but that in the field the new directions were accepted rather slowly.

Clearcutting had become increasingly controversial throughout the 1960s, and the tempo picked up. It was obvious to many that logging practices on the Monongahela and Bitterroot National Forests were not in the best interest of the agency or the environment. On-the-ground review, which included research personnel, recommended modifications, but it was too late.

Litigation in West Virginia and local opposition led by the dean of forestry at the University of Montana boosted clearcutting practices to national prominence. Silviculture specialist Carl Ostrom, who was director of the Division of Forest Management Research, accepted the challenge of preparing a scientific report—*Silvicultural Systems for the Major Forest Types of the United States*—that characterized clearcutting as an acceptable practice under appropriate conditions. In the introduction he wrote that "great public interest now focuses on the harvest cutting system." The manager must make plans in the "context of a complete system of forest culture." Trees were an important part of the calculus but so too were wildlife requirements, the trade-offs for using prescribed burning, and of course social values. No doubt this firm science base was instrumental in having clearcutting permitted under the National Forest Management Act of 1976, itself a spinoff of the Monongahela litigation.

Shortly after Arnold had returned to Washington, the General Accounting Office issued a report that was critical of Forest Service research for failure to convert research findings to field practice. This report put additional emphasis on application and led to the placement of an assistant director of planning and application at each experiment station and the Forest Products Laboratory. In this same way, involvement with the clearcutting controversy through vehicles such as Ostrom's report offers a good example of just how research can make a difference.

A Quality Environment

In 1972 the Forest Service released *National Forests in a Quality Environment*, a fifty-one page action plan in response to the National Environmental Policy Act. It grew out of the internal studies of logging practices that the chief had ordered in 1970. Ostrom provided the silvicultural material. He knew that there were "obvious things that needed correction." Not only were some clearcuts too big, they were not all well designed. In many cases, the local ranger had neglected to follow through with standard procedures to obtain satisfactory regeneration. The report contained specific assignments for the research and administrative branches. The studies identified thirty problem areas, and the chief then formed a task force to articulate appropriate responses, which were then reviewed in the field. A 1972 publication resulted from that effort, and it offers a candid appraisal of the situation; there must not be any more Bitterroots or Monongahelas.

The publication singled out problem number 30 for special attention, calling for an accelerated research program in three areas. These were: understanding the forest as the essential environmental and productive base for multi-resource management; understanding human interactions with the forest environment; and developing timber management practices that enhance or have minimum impact on the environment and are compatible with other forest uses. Problem 30, and the preceding twenty-nine, gave a clear shape to certain portions of the agency's research mission, in coordination with its management arm. Not only was this a tangible response to environmental realities, it was at least in part a response to the GAO report that was critical of the limited amount of technology transfer. Congress was an important player, as there was a definite funding shift from basic research toward applied research and development. This shift received strong support from the White House.

Changing Times

The times were surely changing; shifts in language in only a year or two were revealing. What had been studies in watershed protection were now studies of "degradation" caused by improper pesticide applications and siltation from the construction of logging roads. And the program itself was called Watershed and Aquatic Habitat Research, a broader concept than the previous Watershed Research. It was the same for wildlife studies, those in Wildlife Habitat Ecology Research reported on their investigations related to song birds, endangered species, and other nongame wildlife. Of course, endangered species had been a concern for some time, but it was the shift of emphasis. Of the 109 endangered species listed in 1973, 38 were found on or near national forests.

Yet another sign of the changed times was the announcement that EPA had approved Zectran, the pesticide that the Forest Service saw as its best substitute for DDT after screening 130 chemicals over six years. EPA admin-

Forest Service Chief John McGuire (right) chats with Brock Evans, Washington, D.C. representative of the Sierra Club, March 1973. (U.S. Forest Service photo.)

istrator Russell Train had experienced unprecedented and relentless pressure from Oregon's congressional delegation as the Forest Service sought permission to use DDT to control the tussock moth in the eastern part of the state. Permission was reluctantly granted, but under the most stringent monitoring procedures that could be devised. Forest Service staffers remember preparing a "seven pound" environmental impact statement as part of the approval process. The 1973 application of DDT would be its last domestic use.

As always the Forest Products Laboratory had much to report. They had teamed a computer with a sawmill in a way to extract 10 percent more lumber from each log. The process was dubbed Best Opening Face, and it was aimed at the much smaller and second growth logs being milled. Given that in 1970 there were fifteen billion board feet of such logs being manufactured into lumber, the potential was easy to see. In cooperation with the Department of Housing and Urban Development, the Lab continued to design low-cost housing and had five plans available. Throughout, the Lab sought better ways to utilize residue, the bits and pieces left over when a round log was converted into square material. Project STRETCH used ultrasonic sound to detect defects in logs before they were sawn, allowing the operator to make strategic adjustments and make the most efficient cut.

Activities at Idaho's Priest River Experimental Forest offer a good example of the varied nature of specific research projects at this time. There were per-

R. Keith Arnold (standing) and Robert Buckman present the Forest Products Laboratory's designs for low-cost homes to a women's group. (Forest History Society photo.)

manent growth plots to be remeasured, as were the treated and untreated pairs of plots for a fertilization study. Then there was the collaborative effort as part of the International Biological Program, where the forest provided a coniferous biome coordinating site. Investigators were looking at water from various angles—calibrating a watershed for a new study, analyzing data continuously collected in another watershed for more than thirty years, and calculating snow interception and losses through evaporation in different types of forest stands. Then there were the seven genetics studies and likewise seven for tree diseases. Finally, scientists from the Northern Forest Fire Laboratory in Missoula had run test fires in fuel beds constructed on the forest.

Of course there were many other examples that suggest the range of activities. Pathologist George Hepting published his *Diseases of Forest and Shade Trees of the United States* in 1971, and the following year Peter Koch published *Utilization of the Southern Pines*. Later, Koch would publish in three volumes *Utilization of Hardwoods Growing on Southern Pine Sites*. At the Rocky Mountain Station, Jerome S. Horton produced a 192-page bibliography on the relation of water to vegetation management, while at the Northeastern Station Norman R. Dubois was deeply immersed in his long term investigations of biological control of insects, especially the gypsy moth. In the Pacific Northwest, entomologist Robert Furniss, who many felt was the successor to Paul Keen, collaborated with Val Carolin on a project that would lead to the publication of *Insect Enemies of Western Forests*.

On occasion, a scientist would receive a distinguished honor. Two notable

67

examples were an accumulation of achievement by Don Marx during his long involvement with mycorrhizal research and Kent Kirk's contribution to the understanding of the enzymatic relationships that cause lignin to decompose. They each received the Marcus Wallenberg Prize that carries a one million Swedish kronor stipend. The scientists had received recognition from the Forest Service, as well; Marx's laboratory was renamed as the Institute for Mycorrhizal Research, and for Kirk there was the Institute for Microbial and Biochemical Technology. Kirk would also be elected to the National Academy of Sciences.

Science was moving ahead, but station organization continued to be a pressing problem. We have seen the earlier shift to research centers and the subsequent shift to project organization. Budgets had grown rapidly, bringing more personnel and facilities. But the growth was not uniform; some areas such as timber management and insects and disease had grown more rapidly than some of the others. Too, leadership workload varied widely between and within stations. An inhouse study proposed that the administrative organization based upon subject matter be scrapped and replaced by a system of deputy and assistant directors to deal across disciplines with support services, planning and application, and continuing research.

Station directors were skeptical in that specialists would be reporting to nonspecialists, but Arnold recommended that the study recommendations be adopted. Chief John McGuire agreed, but with the condition that stations be given four years to achieve full implementation. The process sorely tested Arnold's administrative skills, and even then some stations did not fully comply. Years later, some staff still carried ill feelings about the imposed changes, a condition that Arnold's successor would inherit.

Dickerman Cools Things Off

Keith Arnold left the Forest Service abruptly in 1973 to accept a senior research position at the University of Texas. Chief McGuire asked the associate deputy, M.B. Dickerman, to move up. He also asked Dickerman to let the research program cool off after becoming superheated under Arnold. Through its long experience with decentralized management, the agency well understood that its real work was accomplished in the field and that too much horsepower in Washington could lead to frustration as the bench scientists strived to keep up with the changes. It was a time to consolidate substantial gains. Although not obvious to all at the time, it was also the beginning of tighter budgets following the booming growth of the 1960s.

Dickerman was eminently qualified to be deputy chief for research, as he neared the end of his long Forest Service career. After fourteen years as director of the Lake States Experiment Station, he moved to Washington to join a special USDA team to develop a ten-year departmental research plan for Congress to use as an appropriations guide. For several years the Forest Service used this study to justify research proposals. On completion of this assignment, he was asked to be George Jemison's associate deputy; when he declined to be Jemison's successor, he stayed on as Arnold's associate deputy. Arnold recalled how much easier his job had been because of Dickerman's penchant for a "clean desk." The associate handled the day-to-day administrative routine with quiet efficiency.

Dickerman's workday began with a meeting with the chief and other deputies. Meetings of chief and staff dated back to the agency's origin, but during the early decades they had been weekly. By the 1950s issues were evolving so rapidly that the meetings were held more frequently. By the 1960s the chief and staff met every day, with few exceptions, for fifteen minutes up to a couple of hours. To some extent, it was "show and tell," as each deputy reported on the day's pressing issues. Obviously, the chief and the deputies for national forest administration and programs and legislation felt the brunt of most controversies, such as clearcutting, while the deputy for research faced few emergencies. Additionally, senior personnel decisions agency-wide were made at these meetings; in sum, the deputy chief for research was well briefed in all elements of the Forest Service mission.

Dickerman applied his quiet efficiency to finally resolving an issue that had been around for a long time. The Madison lab, while widely supported and admired, did not fit all that efficiently into the agency's administration. In sum, it did things its own way. Since Harper's time, research leadership strove for better integration. For example when the experiment stations reorganized to improve "cross fertilization" among scientists, the lab declined to go along and maintained the pyramid style of organization. Authority was strictly delegated along functional lines. Shifts in personnel helped, and Jemison told director Fleisher, "Herb, when you get back to Madison, we want you to make the Laboratory part of the Forest Service." Arnold continued the pressure, followed by Dickerman's directive that the Lab was to use the same reporting and planning procedures and organization as did the experiment stations. Integration was also enhanced by assigning tours of duty at experiment stations for senior Lab scientists.

At another Forest Service laboratory, the fire lab at Riverside, work was well along in developing a computer-driven management information system via Project FIRESCOPE. Contractors with background in the nation's aerospace program conducted much of the research. FIRESCOPE developed the "methods and vocabulary" for interagency cooperation in fire suppression. This included the Incident Command System, used worldwide to manage situations where groups of people worked in complex emergency situations, such as fire, floods, and earthquakes. Observers rate FIRESCOPE as one of the agency's most productive research and application programs.

In 1974 Congress passed the Forest and Rangeland Renewable Resources Planning Act, RPA for short. The law mandated a beefing up of the agency's already substantial planning process. There was to be an assessment of all forest lands in the United States, followed by detailed programs for the national forest portion of the nation's forest land base. The research branch contributed importantly to the assessment process, because of its skill in collecting and interpreting data, and the program side often required new research to figure out the best way to achieve a new goal. Especially significant, the agency and the Congress were coordinated in developing plans that covered long time spans; the theory was, funding from Congress would also be coordinated and thus more predictable. Subsequent studies show that the theory was borne out in practice, that Forest Service budgets for RPA programs in general and for research in particular increased more rapidly after RPA, at least until the 1980s.

Dickerman was especially pleased with the 3-Bug Program. Many insect pests had plagued American forests over the years, but three species in particular were not only troublesome but persistent. Needed was a large, coordinated, and accelerated program that would capture the interest of Forest Service investigators, as well as its university and state agricultural experiment station collaborators and Congress. The coordinated 3-Bug Program was focused on the gypsy moth in the Northeast, the southern pine bark beetle in the South, and the tussock moth in the Pacific Northwest. Congress appropriated $6 million for the program at a time that the whole research budget was just over $70 million, an impressive increase that reflects the broad base of support that the 3-Bug Program had. The effort was one of the most efficient and effective gathering and interpreting of a vast array of complex information that Dickerman had seen. Similar, multiple projects had been attempted earlier, but none had "held together" the way that the 3-Bug Program had. It could serve as a model of what could be done. The final result was a comprehensive report in book form for each the the three insects studied.

Technology transfer—how to have more of it—remained high on the agency's agenda: "The full benefit of research is never realized until it has been put into use." The official nomenclature was RD&A—research, development and application. Examples were the 3-Bug Program, CANUSA where the United States and Canada would team up to tackle the spruce budworm, and SEAM, the acronym for surface, environment, and management.

SEAM, borne out of the energy crisis of the early 1970s, was aimed at the reclamation of coal strip mines, which researchers had studied since the late 1930s. SEAM began its official life in 1973 through the amalgamation of on-going projects that involved eighteen universities and eight Forest Service research units. The earlier work on "spoils banks" had been centered in the old Central States Station. In the nine states for which it was responsible, the investigators had estimated that there were one hundred thirty thousand acres that had been strip mined. Of interest to other stations was coal mining in West Virginia, Pennsylvania, and Alabama. There was also strip mining for phosphates in Florida and iron ore in Alabama and Minnesota.

Forest Service studies focussed on ways to reforest the areas, which obviously varied a great deal in terms of soil, climate, and the general conditions following cessation of mining. Sample plots tested a range of hardwood and softwood species in order to determine survival rates. As with the later SEAM, there was much involvement with universities, other agencies, and the mining companies themselves. The work of SEAM, plus strengthened state and federal laws, largely quieted the fierce controversies surrounding surface mining in the 1970s.

The Buckman Administration

Dickerman retired in 1975, after serving three years as deputy chief. For his successor, Chief McGuire named Robert E. Buckman, who had been director of the Pacific Northwest Forest and Range Experiment Station. In 1975 Buckman moved to Washington to be associate deputy under Dickerman; much earlier he had worked for Dickerman at the Lake States Station, and he had held a variety of assignments in Washington during the mid-1960s. Buckman served as deputy chief for research for ten years, equal to the combined total of his three predecessors. Staff numbers would vary from year to year, as they had since Harper, but at the beginning of his administration he directed up to a thousand scientists spread among eight experiment stations, the Forest Products Laboratory, Institute of Tropical Forestry, eighty-one field locations, and ninety-three experimental forests and ranges. By the end of his tenure, budget reductions were a fact of life that his successors would inherit.

The Agricultural Research Policy Advisory Committee (ARPAC) was the top level USDA planning committee, and the deputy chief for research as well as a representative of the McIntire-Stennis forestry schools were members. At a meeting shortly after he became deputy, Buckman listened as others spoke favorably about an agricultural conference recently completed in Kansas City, which had sampled widely among agricultural users. ARPAC cochairs Orville Bently and assistant secretary Robert Long urged forestry research to adopt this approach. Buckman and Missouri forestry dean Donald Duncan, the McIntire-Stennis representative, thought that it offered a good model for them to emulate. They organized four regional workshops attended by two thousand people from universities and other agencies to suggest research priorities. In Buckman's mind, they were "laying out a program for Forest Service research for RPA" and for university research under McIntire-Stennis.

The effort was brought together during three days in January 1978, when one hundred delegates from government, professional, environmental, and consumer groups convened in Washington "to identify and discuss national and international research needs and priorities." The cosponsors with the Forest Service were the Cooperative State Research Service, which administered the McIntire-Stennis program, and the Association of State College and University Forestry Research Organizations, which consisted of the sixty-one schools in the McIntire-Stennis program.

Research was to be cooperative and where appropriate it was to be interdisciplinary. It was also to be useful; it was the "A" in RD&A or it might be called "technology transfer" or sometimes "applied research." By whatever label the fifteen hundred to two thousand annually published research reports were to be useful. The publications include major definitive monographs that remained the authority for a particular topic for years, as well as arcane treatises read only by specialists. Overall, Buckman believed that the true contribution was the continuous accumulation of knowledge, the myriad bits and pieces of information, "building blocks" of monographs, articles, papers, and reports that when added together equalled an impressive sum.

Another kind of impressive sum stemmed from an analysis by Robert Z. Callaham, who was director of forest environment research in the Washington office. He identified eighty-one innovations, such as improved utilization, that had stemmed from Forest Service research during fiscal years 1977 to 1979. Of that total he calculated that twenty-two contributed direct monetary benefits of $2.6 billion. This impressive sum would pay the cost of the entire Forest Service effort for the next twenty-four years, based upon the 1979 research budget. A later study estimated the return on oriented strand board research was in the neighborhood of 20 percent, while containerized seedling studies could potentially yield a whopping 80 to 110 percent. Research was yielding a high rate of return. It was useful.

A New Research Act

The McSweeney-McNary Act of 1928 had served Forest Service research well, but a full half-century had passed; many felt that more specific authorizations would be useful. The Forest and Rangeland Renewable Resources Research Act of 1978 resulted from a solid coalition in support, following passage of the 1976 National Forest Management Act. Senator Talmadge perceived the 1976 act to be "western" in that was where most of the national forests lay, and he wanted something more specifically for the South. Cooperative forestry and research, he felt, would benefit the South. During the legislative process, research received its own bill. In 1981 when the agency prepared briefing material for the incoming Reagan administration, the 1978 research act was listed among the five "basic legal authorities" for the Forest Service.

The act divided Forest Service research into seven broad categories; most of them, like timber management research, were familiar. Added was specific authority for the international program, which former chief R. Max Peterson remembered had been "tweaked" in for clarification. The act articulated multiple use research, which included outdoor recreation and other topics that were not well defined in the earlier law. To Buckman, "we attempted to anticipate as many as possible the areas of inquiry that forestry and renewable natural resources might one day address."

Change of Administration

"Draconian" is one of Buckman's favorite adjectives when discussing the budget situation following Ronald Reagan's election. Under the preceding Carter Administration, research had done well in terms of budget, but that quickly changed. Budgets that only kept pace with inflation but at the same time had to be responsive to new demands to study environmentally sensitive topics such as atmospheric deposition and to hire new employees under affirmative action requirements meant reductions in current personnel, between

5 and 10 percent. Buckman would call the station directors to Washington in late December—they called the meetings "Christmas parties"—and they would go through the research program project by project to determine which scientists would be let go. They adopted uniform criteria with which to evaluate the projects, in order to maintain objectivity. Members of Congress are always concerned about job losses in their districts, and the uniform application of the criteria made it easier to defend the difficult and sensitive personnel decisions. There was a plus or two; a few unproductive units were terminated more easily than otherwise and the process generated some discretionary funds that could be used to support new programs, such as biotechnology and economic evaluation of research.

Assistant secretary John Crowell believed that the research budget should be reduced from $130 million down to $100 million, as he saw the benefits to be too far in the future. Chief Peterson and Buckman invited Crowell to tour the Forest Products Laboratory to see for himself that much of the work was applied and high payoff in nature. But Crowell was not convinced and later sent Peterson a "stinging memo" for requesting an "excessive" budget. But Congress was more supportive of forestry research than was the Reagan Administration, and the cuts included in the president's budget were generally restored. The research budget bottomed-out in 1983 just below $110 million and by 1985 was $121 million. The increases would continue but were always tempered by inflationary erosion.

Grants, whereby scientists could apply for support from the USDA Competitive Research Grants Office, were authorized by the 1978 Research Act. The program had lost its funding during the Reagan years, but by 1985 the grants had been restored to a new high of $7.8 million. All were eligible to apply—federal, state, university, or industrial investigators. It was highly competitive with perhaps 10 percent being successful. The fifty-four awards for 1985 averaged $135 thousand for a three-year period; twenty-eight awards supported harvesting, processing, and utilization, while twenty-six treated forest biology and biotechnology.

There was another way to trim federal outlays, and that was to engage in more cooperative research. The W. R. Grace Commission wanted to know if the forest products industry supported research at the Forest Products Laboratory; after all, the industry was receiving direct benefit. The upshot of it was, the Lab was placed under great pressure to augment its reduced budget through substantial cooperative research with the private sector. Director Robert Youngs announced this policy shift during the 1983 meeting with its industry liaison committee. The industries that benefited were asked to share the costs. At some tension during these collaborative efforts was the industry's proprietary interest in specific investigations; where it was likely that a company would intend to capture the benefits, then it would be better to go it alone. Such interests tended to be on the products side of research, rather than on the forestry side.

Biotechnology

Biotechnology was one of the newer research areas, and it held great potential. By discovering an enzyme that could break down lignin, one of wood's primary constituents, the door was opened to new processes in pulping, wood processing, bleaching, and cleaning mill effluent. And these processes did not require chemicals and invasive contaminants or expensive mechanical means. The enzyme could achieve the goal "naturally." Biotechnology also included manipulating DNA or using genetic selection to develop disease-resistant and insect-resistant strains of trees. Rachel Carson had warned against the heavy applications of insecticides that altered the environment so abruptly that normal evolutionary responses were being overwhelmed. So too had international commerce introduced blister rust and chestnut blight that had been unknown to the American ecosystem, with devastating results. These pathogens were now endemic, and biotechnology had the potential of compressing a millennia of evolution into a decade. The American chestnut, once 40 percent of the Appalachian biomass and now close to extinction, could be brought back. The story is less dramatic with white pine's battle with blister rust, but here too recovery was in the offing.

Atmospheric deposition research expanded greatly in 1985, in part through an infusion of funds from the Environmental Protection Agency. Sulfur dioxide emissions from the combustion of fossil fuels was returning to earth as "acid rain" and causing the acidification of aquatic regimes. Researchers undertook to determine the effect, no mean task when there were other simultaneous impacts to filter out. Were observed changes due to acidification or something else? Forest Service scientists were sucessful in having the studies expanded well beyond acid rain. Early results contained few clear answers, but it seemed that some aquatic systems and some vegetational types were much more sensitive than others. As so often in nature, there is no single cause, no single effect, no single solution.

The Illusive Ivory Tower

Acid rain, like clearcutting and other forestry-related issues, was becoming politicized. Electrical generation in Ohio was affecting the forests and lakes of Quebec, and there were international factors to consider, in addition to those raised as part of an ever-increasing concern about environmental quality. In contentious situations, detached investigators could become distracted, as pressure was brought to bear to find the "answer." Sometimes it mattered how the initial question was phrased; entomologists trying to curb the southern pine beetle by removing the older trees that were most susceptible might find themselves in conflict with those studying the endangered red cockaded woodpecker, which preferred to nest in those same trees.

In the Pacific Northwest there was a major conference to bring together all known information on the northern spotted owl, a species that by 1985 had yet to be listed as either threatened or endangered but clearly held the potential for a significant impact on forest practices. Many of the questions dealt with matters of science—how the plant or animal species related to its habitat. But there was another series of questions that was policy-driven—and to that extent politicized—where managers had to make decisions that affected both the habitat and the species that depended upon it. There had yet to be an ivory tower for Forest Service research, and it was clear that there never would be.

International Program

As we have seen, the Forest Service became officially involved in foreign or international forestry at the close of World War II, largely via the FAO forestry program. Another overseas outlet was IUFRO, with Harper serving a term as vice president and later with Jemison as president. Under Jemison, the 1971 World Congress convened in Gainesville, Florida, the first time that the international body had met outside of Europe. Arnold and Dickerman were also active in IUFRO, both serving on its Executive Board. Then there was involvement through USAID, UNESCO, International Society of Tropical Foresters, the International Union of Societies of Foresters, the North American Forestry Commission, and other domestic and foreign institutions and agencies.

Other than these activities, for many years the agency's sustained, on-the-ground international research involvement was that of Frank H. Wadsworth, who since 1942 had been operating out of the Forest Service research facility in Puerto Rico. It was Wadsworth who would go to Panama in cooperation with a USAID watershed project, or to Borneo to examine timber harvesting methods, or to wherever there was a need. Wadsworth, who coincidentally had shared an Ann Arbor office with Arnold as they earned their Ph.D. degrees at the University of Michigan, became "Mr. International Forestry." For many years, Wadsworth reported directly to the chief, a strange situation when it was essentially a one-man program. Eventually administration was transferred to the Southern Experiment Station. The Puerto Rican research program grew, and in 1962 was renamed as the Institute of Tropical Forestry, in part because Harper saw that it was so heavily involved with training foreign nationals that it merited a name that reflected what it did.

Buckman, too, was an advocate of a strong international program for the research branch. He was also active in IUFRO, serving both on the Executive Board and as president. The research act of 1978 had explicitly authorized international forestry, which had suffered from fluctuating emphasis caused by being tied more to American foreign policy than forest policy. It was at a low ebb, down to about six people, when Buckman became deputy, and he aimed to strengthen it. The biggest boost would come through an agreement with USAID for the Forestry Support Program that Deputy Chief Thomas Nelson had orchestrated. The USAID money paid for Forest Service staff to develop skill rosters that could be called upon for forestry programs worldwide. With funding from other sources as well, by 1986 International Forestry would have a staff of twenty-six.

In 1991 International Forestry was split off from Research and placed under its own deputy chief. The Institute of Tropical Forestry was renamed as the International Institute of Tropical Forestry, reflecting the new emphasis. Scientists would of course continue to provide expertise, as would Forest Service personnel from the other branches, but International Forestry now had its own place at the table, but only briefly. In 1995 as President Clinton and Congress pared the federal budget, International Forestry as an entity was "defunded." Much of the work would continue but at reduced levels and as part of other agency programs.

Research Independence

Soon after Peterson became chief in 1979, he combined programs for the national forests with those for State and Private Forestry in Region 8, headquartered in Atlanta. The new organization streamlined management and saved some money through personnel reduction. As it had since Clapp's time,

Research continued to report directly to the chief. Peterson received a suggestion from the state forester of Alabama that research in the southern region also be placed under the regional forester. Not only would such a reorganization save additional money, C.W. Moody stated, but it would "make research more effective by putting it closer to the user group. Getting research results to the user and getting feedback on problem areas and priority of needs should be quicker and easier." Peterson shared the letter with Buckman; following a discussion on how best to respond, the deputy drafted a letter for the chief's consideration.

The issue was not a minor one and had been a concern since Clapp's time. Buckman would later assert that never had research under his administration been prevented from conducting a particular study, nor asked to withhold information, nor had the findings of his scientists been challenged because of a conflict with policy. Research was in fact independent, although there were the occasional grumblings from field officers when an investigator's independence had more the appearance of indifference. At least the scientist might show a little more sympathy with the plight of those working on the "crisis of the day." Some might wonder, however, if research that proposed to lead into areas that might conflict with policy would be more difficult to fund.

Buckman was especially pleased that before Peterson approved his draft that explained to the state forester the historical separation of research, the chief had added a single sentence: "experience has indicated that if action and research programs are combined, action programs will inevitably capture research." Buckman circulated the chief's letter to the station directors and staff, "the issue of the independence of research within the Forest Service was put to rest" for that period at least.

More with Less

Buckman retired in 1986, and was succeeded by John H. Ohman, who had been deputy chief for state and private forestry. Earlier, Ohman had spent many years at the North Central Experiment Station, had held key research administration positions including being Buckman's associate deputy, and was familiar with the breadth and depth of Forest Service research.

One of his first tasks as deputy was to reacquaint himself with research programs, and he made a number of trips to the stations for a quick overview. He found "the enthusiasm and dedication" of the field scientists to be remarkable, in the face of shrinking budgets. Ohman was infected by this enthusiasm and later could see that it helped him cope with the "budget-cutting mania" that was becoming a standard feature of federal service.

During Ohman's tenure, sensitive "interfaces" were receiving continued scrutiny, those between forest and atmosphere, between wildland and urban areas, as well as between wildlife habitat and timber management practices. Studies on international trade and biotechnology, as well as establishment of additional research natural areas were reported.

Ecosystem Research

Although the President's budget would not include a line item for "Ecosystem Research" until 1993, the ground was well laid during the Ohman and earlier administrations, with Forest Service research continuing to produce the knowledge base to implement ecologically-based forest management. With a balanced program of fundamental, applied, and developmental research, the program was responsive to the rapidly changing forestry scene. Programs and Assessments of the Resources Planning Act generated need for expanding research, as had the National Forest Management Act of 1976 with its mandate to sustain both biological diversity and soil productivity. The Endangered Species Act and Clean Water Act also added a sense of urgency to their related corners of research. At the same time there would be a shift in how that land would be managed, as the agency moved away from operating along functional lines toward ecosystem management via an interim process called New Perspectives.

Ohman retired in 1989, and Chief Dale F. Robertson appointed Jerry A. Sesco as the next deputy chief for research. Sesco began his Forest Service career in the National Forest System in 1963 but moved into research that same year. Prior to becoming deputy, he had been a project leader, an assistant station director, a staff assistant in the Washington Office, a station director, and Ohman's associate deputy. In 1997 Chief Mike Dombeck asked Sesco to be his special assistant, and named Robert Lewis to be deputy chief for research. Lewis had been director of the Northeastern Station.

Strategy for the 90's

During his tenure, Sesco implemented a new approach to research management and leadership. He established the "Green Team," an executive group consisting of the station and staff directors, to oversee a broader and more consensus-driven management. He gave increased emphasis to recognizing and rewarding scientists for outstanding contributions. He also implemented a revised peer evaluation process that rewarded collaborative and team search efforts and gave more credit for the significance and impact of research conducted. Workforce diversity, which George Jemison a generation earlier had seen as one of his most difficult tasks, retained its high priority. Sesco named Barbara Weber director of the Pacific Southwest Station, the first woman to hold a station directorship. She next moved up to be his associate deputy. Likewise he appointed Robert Lewis as director of the Northeastern Station, the agency's first black at that level.

Ohman had commissioned a thorough review of the Forest Service research organization, and Sesco's immediate task was to decide how to respond to the report. Major problem areas identified were deteriorating relationships with universities and other cooperators, lack of visibility in the larger science community which as early as Harper's tenure had been seen as an issue, lack of a clear mission, and overall reluctance to change. Sesco's most important follow-up action was development of "Strategy for the 90's." The strategy placed high emphasis on understanding ecosystems, the relationship between people and those ecosystems, and resource options.

In September of his first year as deputy, Sesco described the Forest Service research program to a congressional committee. The agency's goal was to "develop the knowledge and the technology needed to increase environmental and economic values of all the 1.6 billion acres of forests and associated rangelands." He characterized their research as being 40 percent basic and 60 percent applied and developmental. During the past decade the research appropriation had increased from $109 million to $138 million, but its purchasing power had declined. Softer budgets had caused a steady reduction in personnel, and during the same period the number of scientists went from 964 to 720. There was a parallel closing or consolidation of projects and laboratories. For example, the three fire laboratories were merged into the Missoula program, and the Southern Station in New Orleans merged into the Southeastern Station in Asheville. As had Buckman and Ohman, Sesco would have to live in a smaller world, even though the RPA Program from 1975 onward insisted that more research was necessary. Some of the new information was to come from shifts in emphasis—a greater proportion of effort was to study recreation, wildlife, fisheries, and water and to enhance the "compatibility of competing resources."

A Sampling of Research

In 1992 the agency published a pamphlet, Forest Service Leaders in Conservation Research. The attractive publication offered glimpses of ten scientists to represent the diversity of approaches. Ariel Lugo in Puerto Rico was studying many aspects of tropical forests. Thomas Crow in Rhinelander was investigating landscape fragmentation and maintenance of biological diversity. Sue Conard was monitoring fire effects in vegetation at the San Dimas Experimental Forest. At the Forest Products Laboratory, Kent Kirk sought practical applications for wood decay mechanisms. At the Shrub Sciences Laboratories in Utah, Durant McArthur developed hybrid sagebrush for cattle food. Geologist Fred Swanson in Oregon's H. J. Andrews Experimental Forest related natural disturbances to management needs. Linda Joyce at Fort Collins helped to develop mathematical models to forecast forage supply. At La Grande, Jack Ward Thomas organized and directed wildlife task forces to examine sensitive issues. In North Carolina at Coweeta, Wayne Swank conducted long term watershed studies. Also in North Carolina, geneticist Gene Namkoong advocated maintenance of genetic diversity. That same year, the Forest Service reported 2,652 "research accomplishments"—books, articles, reports, and information in a variety of forms by its more than seven hundred scientists. No short sample could do justice.

As the above sample suggests, substantial ecosystem research was being conducted. Long term monitoring was an important constituent of this effort, which included studies related to global climate change and forest health. The forest health program was initiated in 1990 to develop baseline data; continuous monitoring would then allow a reliable measurement of changes in forest conditions brought about by both natural and human-induced events. As so often happens in research, studies overlap, and information gained by investigating climate change was of direct interest to wildlife biologists studying the threatened grizzly bear in the Yellowstone region. As they learned, pine nuts from the whitebark pine were an essential part of the grizzly diet. And the whitebark pine is a high-elevation species, growing in an especially fragile ecosystem that could be altered substantially if the climate warmed. The building blocks continued to add up.

The Owl and the Woodpecker

Researchers on endangered species looked at diverse situations. Hurricane Hugo had severely damaged the Francis Marion National Forest in South

81

Carolina, which held the primary population of the endangered red cockaded woodpecker. The storm destroyed 87 percent of the cavity trees and killed 60 percent of the woodpecker population. Biologists at the Southeastern Station had previously developed artificial nesting cavities, which were installed on the Francis Marion. They estimated that the cavities would allow the woodpecker population to quickly double.

The spotted owl, too, was receiving a great deal of attention. The Forest Service managed 71 percent of its northwestern habitat. Investigators at the Pacific Northwest Station had for years been gathering habitat information, such as the abundance and diversity of prey species. The owl needed a large area of old-growth forest to survive—some scientists were saying as much as two thousand acres for each pair of owls. To set aside that much commercially valuable forest would have a substantial economic impact; the agency projected twenty-eight thousand jobs would be lost by the year 2000.

In 1991 the Northeastern and Rocky Mountain stations collaborated with the Fish and Wildlife Service and the University of Wyoming to publish *Forest and Rangeland Birds of the United States*. The work contains life histories of 518 bird species and provides the land manager insights into the increasingly complex nature of integrating avian considerations into their land management responsibilities. Building on research accumulated since the late 1940s in Alaska, also in 1991 the Pacific Northwest Station published *Influences of Forest and Rangeland Management on Salmonoid Fishes and Their Habitats*. Shortly, decline in salmon populations would share the headlines with owls and woodpeckers.

The Forest Service, National Park Service, Bureau of Land Management, and the Fish and Wildlife Service, which is responsible for administering the Endangered Species Act on the lands managed by the other three agencies, formed the Interagency Scientific Committee chaired by Jack Ward Thomas. The group recommended the establishment of habitat conservation areas, which a high-level interagency task force used to develop recommendations for sustaining the owl while minimizing job loss. Thomas' performance caught an eye in the secretary's office, leading to his appointment as Forest Service chief. Thus he joined Clapp, McArdle, and McGuire who had also come out of research to head the agency.

Mandate for Change

Included in P.L. 101-604, (104 Stat.3545) was a request to determine the capability of current forestry research programs. In response, the National Research Council of the National Academy of Sciences issued in 1990 an eighty-four page report, *Forestry Research: A Mandate for Change.* The Council had established a Committee on Forestry Research "to create a vision of what such research must be like in the future in order for society to achieve desired forest management goals." The committee was sponsored by the familiar combination—Forest Service, Cooperative State Research Service, Society of American Foresters, and the National Association of Professional Forestry Schools and Colleges.

The NRC report expanded upon Sesco's congressional testimony, that federal support for forestry research had been in decline for a decade. Inflation had eroded the Forest Service program by 14 percent. At the same time, McIntire-Stennis funding hovered at 13 percent of the Forest Service budget, scarcely one-fifth of the authorized 50 percent. Not only had the decade seen a 25 percent reduction in the number of scientists, but also the number of Forest Service research units had been reduced from 247 to 190.

Conditions in the private sector were parallel; only a dozen of the fifty largest land-owning companies conducted internal research on forest biology, and funding for even that had dropped by 50 percent during the same time period. Forestry research budgets for all sectors could be added up to $350 million, and the Committee on Forestry recommended that it be increased over 40 percent and to $500 million during the next 5 years. The group also outlined a formula by which this ambitious goal might be reached; included was a need for changing funding priorities in the Department of Agriculture.

Too little money was not the only deficiency that demanded correction; forestry research was fragmented between disciplines and between sectors. Needed was an overall advisory policy mechanism, such as the proposed National Forestry Research Council might function. Forest Service research staff, who had been active participants with the Committee on Forestry, prepared an assessment of the report. They described the recommendations as "sound" and meriting support.

The Voyage Begins

Two thousand publications by seven hundred to nearly a thousand scientists, year after year, sounds like a lot, and it is. But considering the enormity and complexity of the forest and rangeland ecosystems of the United States, or just the 8 percent that the Forest Service administers, there would always be more to learn and then apply. One example is with water; the Forest Service manages 128,000 miles of rivers and streams and 2.2 million acres of lakes and ponds. Add to that the winter snow pack in the Cascades, Sierra, Rockies, and other mountain ranges where most of the national forests lay, and where much of the nation's water originates, and the need and opportunity for research are obvious. Division of Forestry chief Fernow had produced scientific reports on water and also testified to congressional committees on the important relationship between forests and water. Congress was persuaded and in 1897 determined that "maintaining favorable conditions of waterflow" was one of the two purposes for which national forests could be established. Wagon Wheel Gap in Colorado, Coweeta in North Carolina, and Hubbard Brook in New Hampshire are just three of the research watersheds that the agency created to continue these lines of investigations.

Similar long lines can be traced for research in products, timber management, and range. Recreation would have to wait until the 1950s to have its own line item in the research budget, and other topics would also have to wait until their time had come, as Forest Service scientists worked to answer conservation's questions. The 1995 Rensselaerville Roundtable that began this story advocated an integration of science and policy making. This echoed Chief Graves' 1918 research policy letter where he had urged the scientists and the administrators to have the same goals. In his foreword to the roundtable report, Chief Thomas wrote that "seeing the results of good science applied on the ground has provided me with great personal satisfaction," an encouraging observation as the agency voyages beyond the maps.

SELECTED REFERENCES

Brandstrom, Axel, Kirkland, Burt, *Selective Timber Management in the Douglas Fir Region*. Charles Lathrop Pack Foundation, 1936.

Briegleb, Philip A., Pechanec, Joseph, "Where We are Now in Forest Research," *Forest Farmer* (XVIII:2) November 1958: 7–9, 28–30.

Buckman, Robert E., "Evolution of Science Policies in the Forest Service." Unpublished, 1969.

Sandra Brown, et al., *Research History and Opportunities in the Luquillo Experimental Forest*. Southern Forest Experiment Station General Technical Report SO-44, 1983.

Caraway, Cleo, *A Forestry Sciences Laboratory . . . and how it grew*. North Central Forest Experiment Station, 1976.

Chapline, William R., "The History of Western Range Research" *Agricultural History* 18(1), 1944: 127–143.

Clapp, Earle H., *Forest Experiment Stations*. USDA Circular 183, 1921.

Clapp, Earle H., *A National Program of Forest Research*. American Tree Association, 1926.

Clapp, Earle H., Unpublished Memoirs. n.d.

Coville, F., *Forest Growth and Sheep Grazing in the Cascade Mountains of Oregon*. Division of Forestry Bulletin No. 15, 1898.

Cowlin, Robert, "Federal Forestry Research in the Pacific Northwest." Unpublished manuscript n.d.

Dana, Samuel T., *Forest and Range Policy*. McGraw-Hill, 1956.

Dickerman, M. B., "History of Forest Service Research." Unpublished manuscript. n.d.

Doig, Ivan, *Early Forestry Research: A History of the Pacific Northwest Forest and Range Experiment Station, 1925–1975*. Pacific Northwest Forest and Range Experiment Station, 1976.

Fairchild, Fred Rogers, *Forest Taxation in the United States*. USDA Miscellaneous Publication No. 218, 1935.

Fernow, B.E., *Forest Influences*. USDA Bulletin No. 7, 1893.

Fernow, B.E., *Report Upon the Forestry Investigations of the U.S. Department of Agriculture, 1877–1898*. H. Doc. 181, 55 Cong 3, 1899.

Geier, Max G., "Forest Service Research in Alaska: A History of Science and Community in the Study of Alaska's Forests." Unpublished manuscript, 1994.

Godfrey, Anthony, "Progress through Wood Research: A History of the U.S. Forest Products Laboratory—Madison, Wisconsin, 1960–1990." Unpublished manuscript, 1990.

Gregoire, Timothy G., "Roots of Forest Inventory in North America." Unpublished paper, Society of American Foresters, 1992.

Harper, V. L., Foreword in *Forest Farmer*. (XVIII:2) November 1958:6.

Harper, V. L., Jemison, G. M., Forsling, C.L., "Early Forest Service Research Administrators." Interview by Elwood R. Maunder, Forest History Society, 1978.

Hough, Franklin B., *Report upon Forestry*. USDA. Vol. I: 1878, Vol. II: 1880, Vol. III: 1882.

Isaac, Leo A., "Douglas Fir Research in the Pacific Northwest, 1920–1956." Interview by Amelia R. Fry, University of California, 1967.

Josephson, H. R., et al., "History of Forest Economics." Interview by Ann Lage, University of California, 1982.

Kaufert, Frank H., and William H. Cummings, *Forestry and Related Research in North America*. Society of American Foresters, 1955.

Keck, Wendell M., *Great Basin Station—Sixty Years of Progress in Range and Watershed Research*. Intermountain Forest and Range Experiment Station Research Paper INT-118, 1972.

LeMaster, Dennis C., *Decade of Change: The Remaking of Forest Service Statutory Authority During the 1970s*. Greenwood Press, 1984.

Lund, Walter H., "Timber Management in the Pacific Northwest Region, 1927–1965." Interview by Amelia R. Fry, University of California, 1967.

Merz, Robert W., *A History of the Central States Forest Experiment Station, 1927–1965*. North Central Forest Experiment Station, 1981.

Munger, Thornton T., "Forest Research in the Northwest." Interview by Amelia R. Fry, University of California, 1967.

Munger, Thornton, T., "Fifty Years of Forest Research in the Pacific Northwest," *Oregon Historical Quarterly* Sept. 1955: 226–247.

National Academy of Sciences, "Forestry Research," 1929.

National Research Council, *Forest Research in the United States*. Processed. 1938.

National Research Council, *Forestry Research: A Mandate for Change*. National Academy Press, 1990.

Nelson, Charles A., *History of the U.S. Forest Products Laboratory, 1910–1963*. Forest Products Laboratory, 1971.

Ostrom, Carl E., "An Interview with Carl E. Ostrom." Harold K. Steen, Forest History Society, 1994.

Price, Raymond, *History of Forest Service Research in the Central and Southern Rocky Mountain Regions, 1908–1975*. Rocky Mountain Forest and Range Experiment Station General Technical Report RM-27, 1976.

Read, Ralph A., "Plains Forestry, 1953–1975." Unpublished manscript, 1975.

Rudolf, Paul O., *History of the Lake States Forest Experiment Station*. North Central Forest Experiment Station, 1985.

Sargent, Charles S., *Report on the Forests of North America*. U.S. Census, 1884.

Schrepfer, Susan R., et al., *A History of the Northeastern Forest Experiment Station*. Northeastern Forest Experiment Station Technical Report NE-7, 1973.

Spencer, John S, Jr., "USDA Forest Service Research and Scientific Publishing: Their

Impact on Forestry Research and Forest Management." Unpublished manuscript, n.d.

Squires, John W., "A Cooperative Research Effort," *Forest Farmer* (XVIII:2) November 1958: 16.

Steen, Harold K., *The U. S. Forest Service: A History*. University of Washington Press, 1976, 1991.

Steen, Harold K., ed., *View From the Top: Forest Service Research by R. Keith Arnold, M.B. Dickerman, and Robert E. Buckman*. Forest History Society, 1994.

Storey, Herbert C., "History of Forest Service Research: Development of a Program." Unpublished draft, 1974.

Train, Russell E., "An Interview with Russell E. Train." Conducted by Harold K. Steen. Forest History Society, 1993.

Tratman, E.E. Russell, *Report on the Use of Metal Railroad Ties and on Preservation Processes and Metal Tie-plates for Wooden Ties*. USDA Bulletin No. 9. 1894.

Trimble, George R., Jr., *A History of the Fernow Experimental Forest and the Parsons Timber and Watershed Laboratory*. Northeastern Forest Experiment Station Technical Report NE-28, 1977.

Wadsworth, Frank H., "The Evolution of Tropical Forestry: Puerto Rico and Beyond." Interview by Harold K. Steen, Forest History Society, 1993

Wellner, Charles A., *Frontiers of Forestry Research—Priest River Experimental Forest, 1911–1976*. Intermountain Forest and Range Experiment Station, 1976.

USDA Forest Service. *Annual Report of the Chief. 1882–1995*.

USDA Forest Service. *Review of Forest Service Investigations*. 2 vols., 1913.

USDA Forest Service. *Half A Century of Research: Fort Valley Experimental Forest, 1908–1958*. Rocky Mountain Forest and Range Experiment Station Paper No. 38. 1958.

USDA Forest Service. *A National Forestry Research Program*. USDA Forest Service, 1961.

USDA Forest Service. *A National Forestry Research Program*. Miscellaneous Publication No. 965, 1964.

USDA Forest Service. *National Forests in a Quality Environment*. 1972.

USDA Forest Service. *Assessment of Forestry Research: A Mandate for Change*. Processed, n.d.

USDA Forest Service. *Navigating into the Future*. 1995.

U.S. Senate, *A National Plan for American Forestry*. 2 Vols. Doc. No. 12, 73 Cong. 1, 1933.

U.S. Senate, *The Western Range*. Doc. 199, 74 Cong 2, 1936.

U.S. Senate. *Timber Depletion, Lumber Prices, Lumber Exports, and Concentrations of Timber Ownership*. Report on S. Res. 311, 66 Cong 2, 1920.

U.S. Senate. *The Forest Service Research Program. Hearings before the subcommittee of the Committee on Appropriations*, 1951.

U.S. Senate, *Agricultural Appropriations. Hearing before subcommittee of the Committee on Appropriations*, 1990.

Zon, Raphael and Sparhawk, William N., *Forest Resources of the World*. 2 vols. McGraw-Hill, 1923.

Zon, Raphael, *Forests and Water in the Light of Scientific Investigation*. GPO 1927.

INDEX

Acheson, Dean, 35
acid rain, 75
Adams, John Quincy, 1–2
Agricultural Research Council, 40–41
Agricultural Research Policy Advisory Committee (ARPAC), 72
agriculture, forestry as part of, 2–3
Aiken, George, 55
air pollution, 60
Alaska Forest Research Center, 46
American Association for the Advancement of Science
    Clapp, Earle H., 11
    Fernow, Bernhard Eduard, 4
    Hough, Franklin B., 2
American Forestry Association, 4
Amherst Experiment Station, 15
Anderson, Mark, 17
Appalachian Council, 19
Appleby, Paul H., 35
applied research. *See* technology transfer.
Arnold, R. Keith
    Branch of Research appointment, 62–63
    clearcutting research, 64–65
    IUFRO involvement, 76
    left Forest Service, 68
    Operation Firestop, 50
    photograph, 63, 67
    projects during administration, 65–68
    recreation research, 63–64
    station organization, 68
    technology transfer, 65
Arnold Arboretum, Harvard, 3
ARPAC (Agricultural Research Policy Advisory Committee), 72
Artificial Regeneration in the Southern Pine Region, 31
Association of State College and University Forestry Research Organizations, 72
atmospheric deposition research, 75

balsa, 33
Bankhead-Jones Farm Tenant Act, 31
bark beetles
    ponderosa pine, 31
    southern pine, 70
Bates, Carlos, 8
Beaver Creek watershed, 51
Benson, Ezra
    photograph, 48
    range research transfer, 44
Bently, Orville, 72
Best Opening Face project, 66
Betts, H.S., 6
Bickford, C. Allen, 47
biological diversity, 81
biological insect control
    appropriations for, 57
    Dubois, Norman R., 67
biometricians added to staff, 47
biotechnology, 75
birds species histories, 82
Bitterroot National Forest, 64
borings, measurements of (photograph), 58
Boundary Waters Canoe Area, 58–59
Branch of Products, 8
Branch of Research
    appointments
        Arnold, R. Keith, 62–63
        Buckman, Robert E., 71
        Clapp, Earle H., 10–11
        Dickerman, Murlyn B., 68
        Forsling, Clarence, 32
        Harper, Verne L., 38–39
        Jemison, George, 57
        Kotok, Edward I., 35–36
        Lewis, Robert, 79

Marsh, Raymond E., 32
Ohman, John H., 78
Sesco, Jerry A., 79
established, 10
Brandstrom, Axel, 25
Briegleb, Philip, 49
Buck, Charles, 26
Buckman, Robert E.
ARPAC involvement, 72
biotechnology, 75
Branch of Research appointment, 71
budget cuts, 73–74
cooperative research, 72, 74–75
Forest and Rangeland Renewable Resources Research Act, 73
international program, 76–77
photograph, 67
politicization of forestry research, 76
research independence, 77–78
retirement, 78
statistical research, 41–42
technology transfer, 72
workshops, 72
"buncombe" quote (Weeks Law testimony), 10
Bureau of Land Management, 35
Bureau of Plant Industry, experimental range transfers, 17
Byrd, Robert, 43

Callaham, Robert Z., 72–73
CANUSA program, 71
Capper, Arthur, 15
Capper Report, 15
Caribbean Forester, 35
Carolin, Val, 67
Carson, Rachel, 57
Cato, influence on forest science, 2
CCC (Civilian Conservation Corps), 30
Central Investigative Committee, 8–9

Chapman, Roy, 28
Charles Lathrop Pack Foundation, 26
Chittenden, Hiram M., 10
CIA-funded translations, 60
cigar box bomb evidence, 30
Civilian Conservation Corps (CCC), 30
Clapp, Earle H.
American Association for the Advancement of Science, 11
appointments
acting chief of Forest Service, 32
associate chief of Forest Service, 32
Branch of Research, 10–11
Capper Report, 15
Copeland Report, 22–24
Depression-era research
budget increases, 22
projects, 29–32
Douglas-fir study, 25–26
experiment stations
implemented, 15–17
proposed, 14–15
Forest Experiment Stations, 14
forest survey
authorized, 21
budget appropriations for, 22
and Capper Report, 15
implemented, 19
rejected by Congress, 14
forest taxation studies, 24–25
list of accomplishments, 24
McSweeney-McNary Act, 19–21
A National Plan for American Forestry, 22
natural areas, 18–19
photograph, 11
range research, 17–18, 26–27
research councils, 19
research independence, 12, 14–15
research natural areas, 18–19
retirement, 32

SAF committee to study forestry
  research, 20
statistical research, 27–29
World War I, 12–14
Clarke-McNary Act, 16
clearcutting
  Arnold, R. Keith, 64–65
  Douglas-fir study, 25–26
Cline, McGarvey, 8
coal strip mine reclamation, 71
Cold War, 51–52
cold-soda process, 52
commercial tree study, 6
Committee Research Evaluation
  (CORE), 41
committees to study forestry research
  CORE, 41
  National Academy of Sciences, 20,
    83
  National Research Council, 20, 28,
    83
  Society of American Foresters, 20
Compton, Wilson, 21
Conard, Sue, 81
coop-aid research, 40
cooperative organization studies, 30
cooperative research
  Buckman, Robert E., 72, 74–75
  Harper, Verne L., 40, 49–50
  Kotok, Edward I., 38
  McIntire-Stennis Law, 55–57
Cooperative State Research Service,
  72
Copeland Report, 22–24
CORE (Committee Research Evalu-
  ation), 41
Cornell University forestry school, 4
Coville, Frederick V., 4–5
Coweeta Experimental Forest, 51, 81,
  84
Cowlin, Robert, 28
Crafts, Edward C., 47
criminal proceedings, evidence, 29–30

Croft, Harold, 51
cross-discipline station organization,
  68
Crow, Thomas, 81
Crowell, John, 74
cultural diversity, 57–58, 80

Dana, Samuel Trask
  Amherst experiment station, 15
  Fort Valley Experiment Station, 7
  Office of Silvics, 8
  photograph, 45
  recreation research, 44
DDT, 37, 57, 66
defense-related research
  Cold War, 51–52
  World War I, 12–14
  World War II, 33–34
degradation studies, 65
Deming, W. Edwards, 27
Demmon, Elwood L., 62
dendrology, 4–6
Depression-era research
  budget increases, 22
  Copeland Report, 22–24
  Douglas-fir study, 25–26
  experiment stations, 30
  Forest Products Laboratory, 29–32
  forest taxation, 24–25
  growth plots, 31–32
  range research, 26–27
  statistical research, 27–29
Dickerman, Murlyn B.
  Branch of Research appointment,
    68
  Forest Products Laboratory inte-
    gration, 69
  IUFRO involvement, 76
  photograph, 62
  retirement, 71
  technology transfer, 71
  3-Bug Program, 70
  War Production Board, 34

Diseases of Forest and Shade Trees of the United States, 67
Division of Agrostology, 5
Division of Botany, 5
Dodds, Gordon B., 10
Douglas-fir study, 25–26, 28–29
Dubois, Norman R., 67
Duncan, Donald, 72
dwarf mistletoe, 57

Eastman, James, 55
ecosystems research
  Arnold, R. Keith, 63–64
  Ohman, John H., 79
  Sesco, Jerry A., 81
Eddy, James G., 30
Eddy Tree Breeding Station, 30
editorial specialists, 53
Egleston, Nathanial, 3
endangered species, 65, 76, 81–82
engineering studies, 43
entomology studies, 43
Environmental Protection Agency, 63, 75
Evans, Brock, 66
Evelyn, John, 2
experiment stations
  federal funds provided (Hatch Act), 5
  first one established, 7
  implemented, 15–17
  proposed, 3, 14–15
experimental forest conversions, 16

Fairchild, Fred Rogers, 25
FAO (Food and Agricultural Organization), 35
federal salary grades, changing, 61–62
Fernow, Bernhard Eduard
  American Association for the Advancement of Science, 4
  American Forestry Association, 4
  "buncombe" quote (Weeks Law testimony), 10
  Cornell University forestry school, 4
  Division of Forestry, 4
  forest influences, 5
  forest/flood study, 10
  pathology studies, 5
  rain-making, 5
  Report upon . . . Forestry Investigations . . . USDA . . . 1877–1898, 4
  timber research, 5
  wood decay, 5
Fire and Water: Scientific Heresy in the Forest Service, 10
fire research
  Conard, Sue, 81
  Gisborne, Harry T., 28–29
  Harper, Verne L., 50
  Jemison, George, 59
  Operation Firestop, 50
  Operation/Project Skyfire, 50, 59
  Project Firescan, 59
  Project FIRESCOPE, 70
Fisher, R.A., 27
Fleisher, Herb, 69
flood research, 9–10. *See also* watershed research.
Food and Agricultural Organization (FAO), 35
Fool Creek watershed, 50
forage studies, 81
Forbes, Reginald, 15
Foreign Forestry Unit, 44
Forest and Rangeland Birds of the United States, 82
Forest and Rangeland Renewable Resources Planning Act (RPA), 70
Forest and Rangeland Renewable Resources Research Act, 73
forest cooperatives, 30–31
Forest Experiment Stations, 14
Forest Farmer, 49
forest/flood study, 9–10

forest genetics, 30, 49, 81

forest influences, 5

forest management, 36, 59–60

forest pathology
    Fernow, Bernhard Eduard, 5
    Harper, Verne L., 43
    Roth, Filibert, 5

forest products, 6. *See also* Forest
    Products Laboratory.

Forest Products Laboratory
    Best Opening Face project, 66
    Cline, McGarvey, 8
    Cold War, 51–52
    criminal proceedings, evidence,
    29–30
    defense-related research
        Cold War, 51–52
        World War I, 13–14
        World War II, 34
    Depression-era research, 29–32
    Dickerman, Murlyn B., 69
    ecosystem research, 81
    established, 8
    Harper, Verne L., 51–52
    Impreg, 52
    industry relations, 30, 74–75
    integration, 69
    kiln-drying research, 36
    low-cost housing, 66
    new building funds appropriated,
    22
    packaging technology, 34
    photographs, 23, 52, 58
    Pinchot, Gifford, 8
    postwar research, 36
    Project STRETCH, 66
    proposed, 8
    World War I, 13–14
    World War II, 34

Forest Research Advisory Com-
    mittee, 48–50

The Forest Research Program, 37–38

Forest Resources of the World, 16

Forest Service Leaders in Conserva-
    tion Research (pamphlet), 81

Forest Service Science and Policy
    Roundtable, 1, 84

forest surveys
    Clapp, Earle H.
        authorized, 21
        budget appropriations for, 22
        and Capper Report, 15
        implemented, 19
        rejected by Congress, 14
    RPA, 70

forest taxation, 16, 24–25

Forest Taxation in the United States,
    24–25

Forest Utilization Service (FUS), 37

Forestry Research: A Mandate for
    Change, 83

forestry schools cooperation, 56–57

Forestry Support Program, 77

Forsling, Clarence
    appointments
        Branch of Research, 32
        Grazing Service, 35
    range research, 26–27
    watershed research, 51
    World War II research, 33–34

Fort Valley Experiment Station, 7, 16

FPL Natural Finish, 52

Francis Marion National Forest,
    81–82

Fraser Experimental Forest, 50

Frothingham, Earl, 15, 23–24

Furniss, Robert, 67

FUS (Forest Utilization Service), 37

GAO, criticism of technology
    transfer, 64–65

genetic diversity, 81

Gila National Forest, photograph, 18

Gisborne, Harry T., 28–29

Glesinger, Egon, 35

glue-laminated construction, 34, 52

Gooding, Jean, 48
Government Employees Training
    Act, 42
grants authorized, 74
Graves, Henry
    Central Investigative Committee
    established, 8–9
    Division of Forestry appointment, 8
    FAO forestry committee, 35
    NAS committee to study forestry
    research, 20
    *Review of Forest Service Investiga-
    tions,* 9
    technology transfer, 9
Gray, Linda, 48
grazing, 4–5, 12. *See also* range re-
    search.
Grazing Service, 27, 35
Greeley, William B.
    National Academy of Sciences
    address, 19–20
    Society of American Foresters
    address, 20
Green Team, 79–80
Grosenbaugh, Louis R., 53
growth plots, 31–32
guayule, 33–34, 37
gypsy moth, 19, 67, 70

habitat studies, 43. *See also* wildlife
    research.
Hague, Arnold, 4
Hall, J. Alfred, 37
Hall, R. Clifford, 25
Hall, William L., 8
hardwood pulp, 52
hardwood research, 19, 36, 49–50,
    59–60
Harper, Verne L.
    administrative issues, 52–53
    Agricultural Research Council,
    40–41
    Alaska, 46
    biometricians, 47
    Branch of Research appointment,
    38–39
    congressional relations, 39–40,
    42–43
    coop-aid research, 40
    cooperative research, 40
    fire research, 50
    Forest Products Laboratory pro-
    jects, 51–52
    Forest Research Advisory
    Committee, 48–50
    forestry research plans, 53–54
    international forestry, 44
    IUFRO involvement, 76
    liaison to National Academy of
    Sciences Research Council, 41
    man-in-the-job concept, 52–53
    McIntire-Stennis Law, 55–57
    Ph.D. degrees, 42
    photograph, 39
    pioneer units proposed, 53
    projects during administration,
    43–47
    range research, 43–44
    recreation research, 44
    retirement, 57
    shelterbelts, 45–46
    statistical research, 27–28
    timber research, 45, 47–48
    university-based forestry research,
    56–57
    USDA relations, 40–42
    War Production Board, 34
    water research, 50–51
    and White House science staff, 54
    Whitten report, 40
Harris, Marguerite, 48
Hatch Act, 5
Hayden, Carl, 43, 50–51
Hepting, George, 67

Herty, Charles, 6–7
H.J. Andrews Experimental Forest, 81
"Hoot Mon" laws, 19
Horton, Jerome S., 67
Hough, Franklin B., 2–3
housing, low-cost, 66
Houston, David F., 10
Hubbard Brook, 84
Hursh, Charles R., 51

Ickes, Harold, 27
Impreg, 52
improved trees, 59
indexes, not allowed, 25
industry relations
    Forest Products Laboratory, 30, 74–75
    Forest Research Advisory Committee, 49
    forest taxation, 25
    McSweeney-McNary Act, 21
Influences . . . on Salmonoid Fishes and Their Habitats, 82
Insect Enemies of Western Forests, 67
insecticides. *See* pesticides.
inspection districts created for national forests, 7
Institute for Microbial and Biochemical Technology, 68
Institute for Mycorrhizal Research, 68
Institute of Northern Forestry, 46
Institute of Tropical Forestry, 77
Interagency Scientific Committee, 82
International Biological Program, 67
International Forestry, 44, 77
International Institute of Tropical Forestry, 77
international program
    Buckman, Robert E., 76–77
    Harper, Verne L., 44
    Jemison, George, 60–61

name change, 44
part of FAO, 34–35
PL-480 monies, 60–61
specific authority for, 73
split off from Research, 77
International Society of Tropical Foresters, 76
International Union of Forestry Research Organizations (IUFRO), 35, 76
International Union of Societies of Foresters, 76
investigations *versus* research, 6
Irion, Harry, 20
Isaac, Leo, 26, 29
(IUFRO) International Union of Forestry Research Organizations, 35, 76
ivory tower, lack of, 26–27, 76

Jardine, James T., 8, 17–18
Jardine, William M., 21
Jemison, George
    administrative issues, 61–62
    Branch of Research appointment, 57
    environmental forestry research, 60
    fire research, 59
    forest management research, 59–60
    international program, 60–61
    IUFRO involvement, 76
    minorities, hiring, 57–58
    Program for the National Forests, 54
    recreation research, 58–59
    retirement, 61
    salary grades, changing, 61–62
    statistical research, 28
    wildlife research, 60
    wind speed gauges, car-mounted, 29
    women, hiring, 58

Jepson, W.L., 6
Johnson, Floyd, 47
Jornada Experimental Range
    transfer from Bureau of Plant Industry, 17
    transfer to ARS, 44
Joyce, Linda, 81

Keen, Paul, 31
Keen Classification System, 31
kiln-drying research, 36
Kirk, Kent, 68, 81
Kirkland, Burt T., 25
Koch, Peter, 53, 67
Koehler, Arthur, 29–30
kok-saghyz, 33–34
Kotok, Edward I.
    Branch of Research appointment, 35–36
    cooperative research, 38
    House Appropriations Committee testimony
        1946, 36
        1950, 37–38
    research centers implemented, 36
    retirement, 38

Lackey, Hendrix, 46
Lake States Experiment Station, 15, 31, 62
Lake States Forest Research Council, 19
laminated beams, 29
landscape fragmentation, 81
larch casebearer, 57
Larsen, Edwin V.H., 53
Larsen, Philip, 53
Lewis, Robert, 79
Lewis, Steven, 48
lightning strike studies, 50
lignin research, 29, 68, 75
Lindbergh kidnapping, 29–30

line-plot method, 28
Little, Elbert, 5
liveoak experiment, 1–2
livestock industry, and range research, 27
Long, George, 21
Long, Robert, 72
low-cost housing, 66
Lugo, Ariel, 81
lumber industry relations
    Douglas-fir studies, 25–26
    Forest Products Laboratory, 30, 74–75
    Forest Research Advisory Committee, 49
    forest taxation, 25
    McSweeney-McNary Act, 21
Lund, Walter, 26

Man and Nature: The Earth as Modified by Human Action, 2
Management of Ponderosa Pine in the Southwest, 45
man-in-the-job concept, 52–53
Marcus Wallenberg Prize, 68
Marsh, George Perkins, 2
Marsh, Raymond E., 26
Marx, Don, 68
Maybeso Experimental Forest, 46
McArdle, Richard E., 48
McArthur, Durant, 81
McGee, W.J., 4
McGuire, John, 66
McIntire, Clifford, 55
McIntire-Stennis Law, 55–57
McNary, Charles, 21
McNary-Woodruff Act, 19
McSweeney, John, 21
McSweeney-McNary Act, 19–21, 46
"micro-wilderness" areas, 59
Mills, Wilbur, 46
minorities, hiring, 57–58, 80

Mohr, Charles, 4
Monongahela National Forest, 64
Moody, C.W., 78
Munger, Thornton
    Alaska, 46
    Douglas-fir study, 26
    Portland experiment station, 16
    research natural areas, 18–19
mycorrhizal research, 68

Namkoong, Gene, 81
National Academy of Sciences
    committees to study forestry re-
    search, 20, 83
    Greeley address, 19–20
    Harper, Verne L., 41
National Environmental Policy Act,
    63
National Forest Management Act, 64,
    73
*National Forests in a Quality Environ-
    ment,* 64
A National Plan for American Forestry,
    22
National Research Council
    committees, 20, 28, 83
natural areas, 18–19
natural disturbances research, 81
naval stores studies, 8, 36. *See also*
    defense-related research.
Nelson, Thomas, 77
Nixon, Richard M., 63
*Nomenclature of the Arborescent Flora
    of the United States,* 4
Norris-Doxey Cooperative Farm For-
    estry Act, 31
North American Forestry Commis-
    sion, 76
Northeastern Council, 19
Northeastern Experiment Station
    biological insect control, 67
    bird species research, 82

cooperative organization studies, 30
    poplar hybrids, 30
Northern Forest Experiment Station,
    46
Northern Forest Fire Laboratory, 67
northern spotted owl, 76, 82
Nutting, Albert, 55

Office of Silvics, 8
Ohman, John H.
    Branch of Research appointment,
    78
    ecosystems research, 79
    retirement, 79
old growth timber stands, 36–37
Operation Firestop, 50
Operation Skyfire, 50, 59
Ostrom, Carl, 46–47, 64–65
Otsego Forest Products Cooperative,
    30
Outdoor Recreation Resources
    Review Commission, 44

Pacific Northwest Experiment Station
    photograph, 16
    renamed Institute of Northern For-
    estry, 46
    salmon studies, 82
packaging technology, 34
paint tests (photograph), 52
pathology studies
    Fernow, Bernhard Eduard, 5
    Harper, Verne L., 43
    Roth, Filibert, 5
Pearson, Gus A.
    Fort Valley Experiment Station, 7
    ponderosa pine studies, 44–45
Pearson, Lester B., 35
Pechanec, Joseph, 49
Perrine, Henry, 1
pesticides
    DDT, 37, 57, 66

2,4-D, 37
2,4,5-T, 37
Zectran, 65–66
Peterson, R. Max, 73, 77–78
Ph.D. degrees, 42
photographs
  Arnold, R. Keith, 63, 67
  Benson, Ezra, 48
  borings, measurements of, 58
  Buckman, Robert E., 67
  Clapp, Earle H., 11
  Dana, Samuel Trask, 45
  Demmon, Elwood L., 62
  Dickerman, Murlyn B., 62
  Evans, Brock, 66
  Forest Products Laboratory
    borings, measurements of, 58
    building, 23
    Lindbergh kidnapping evidence, 30
    paint test panels, 52
  Fort Valley Experiment Station, 7
  Gila National Forest, 18
  Gooding, Jean, 48
  Gray, Linda, 48
  Harper, Verne L., 39
  Harris, Marguerite, 48
  by Jardine, James T., 18
  Lackey, Hendrix, 46
  Lake States Experiment Station
    building, 15
    staff, 62
  Lewis, Steven, 48
  Lindbergh kidnapping evidence, 29
    photograph, 30
  McArdle, Richard E., 48
  McGuire, John, 66
  Mills, Wilbur, 46
  Pacific Northwest Experiment Station, 16
  paint test panels, 52
  by Pearson, Gus A., 7
  pollination research, 59
  Read, Ralph, 46
  seed dissemination, 16
  Sherman, Harold, 46
  Slease Goat Ranch, 18
  Southern Forest Experiment Station, 39
  Starke Branch staff, 39
  Trujilla Creek, 18
  by Varela, A.G., 11
  Zon, Raphael, 62
Pinchot, Gifford
  Division of Forestry appointment, 5–6
  fired by Taft, 8
  forest/flood study, 9–10
  Forest Products Laboratory established, 8
  inspection districts created for national forests, 7
  vision of the agency, 6
  Weeks Law testimony, 10
pioneer units, 53
PL-480 monies, 60–61
Pliny, influence on forest science, 2
politicization of forestry research, 76
pollination research, photograph, 59
ponderosa pine, 6, 44–45
ponderosa pine bark beetle, 31
poplar hybrids, 30
postwar research, 35–38
Potter, Albert, 12
prefabricated housing, 29
Priest Lake Experiment Station, 16
Priest River Experimental Forest, 66–67
Program for the National Forests, 54
Project Firescan, 59
Project FIRESCOPE, 70
Project Skyfire, 50, 59
Project STRETCH, 66
project-based station organization, 53

publications
    editorial specialists hired, 53
    number of, 83
    rules, 25
    translations, CIA-funded, 60
Puerto Rican program, 30, 77. *See also*
    international program.

quinine, 33

rain-making, 5
*Range Management on the National
    Forest,* 17
range research
    Benson, Ezra, 44
    Clapp, Earle H., 17–18, 26–27
    Forsling, Clarence, 26–27
    Harper, Verne L., 43–44
    Ickes, Harold, 27
Read, Ralph, 45–46
Record, Samuel J., 6
recreation research
    Arnold, R. Keith, 63–64
    Dana, Samuel Trask, 44
    Harper, Verne L., 44
    Jemison, George, 58–59
    Lake States Station, 31
    wilderness use, 58–59
    Zon, Raphael, 31
red cockaded woodpecker, 82
regions established for national for-
    ests, 7
Rensselaerville Roundtable (quota-
    tion), 1, 84
Report on the Forests of North America,
    3
A Report Upon Forestry, 3
*Report upon . . . Forestry Investigations
    . . . USDA . . . 1877–1898,* 4
research centers
    discontinued, 53
    implemented, 36

proposed, 36
research councils, 19
research independence
    Buckman, Robert E., 77–78
    Clapp, Earle H., 12, 14–15
    forest/flood study, 10
    Peterson, 77–78
    studies of, 10
research natural areas, 18–19
research *versus* investigations, 6
Rettie, James, 47
Review of Forest Service Investigations,
    9
Robinson, Gordon, 26
Rocky Mountain Forest and Range
    Experiment Station
    bird species research, 82
    water and vegetation management,
    67
    watershed research, 50–51
Roth, Filibert, 4–5
roundheaded pine beetle, 57
RPA (Forest and Rangeland
    Renewable Resources Planning Act),
    70
rubber substitutes, 33–34
Russell, Richard, 40

SAF (Society of American Foresters),
    20
sagebrush, cattle food, 81
salary grades, 61–62
salmon studies, 46, 82
San Dimas Experimental Forest, 81
Santa Catalina Research Natural
    Area, 18
Santa Rita experimental range, 17
Sargent, Charles S., 3
Schiff, Ashley, 10
Schnur, G. Luther, 47
Schumaker, Francis X., 27–28
SEAM (surface, environment, and

management), 71
seed dissemination, photograph, 16
Senior Executive Service, 62
Sesco, Jerry A.
    Branch of Research appointment, 79
    ecosystem research, 81
    endangered species research, 81–82
Shaw, Byron T., 40–41
Shaw, E.W., 53
sheep grazing, 4–5
shelterbelts, 8, 30, 45–46
Sherman, Harold, 46
Shrub Sciences Laboratory, 81
Sierra Club, and Douglas-fir study, 26
Silcox, Ferdinand, 32
Silent Spring, 57
siltation, 65
Silva of North America, 3
silvicultural research, 59–60
Silvicultural Systems for the Major
    Forest Types of the United States, 64
Slease Goat Ranch, photograph, 18
Society of American Foresters (SAF), 20
Southeastern Station, 82
Southern Forest Experiment Station, 39
Southern Forest Fire Research
    Laboratory, 49
Southern Hardwood Forest Research
    Group, 49–50
Southern Institute of Forest Genetics, 49
southern pine, 4, 6–7
southern pine bark beetle, 70
Sparhawk, W.R., 16
spotted owl, 76, 82
spruce research, 19
St. Louis Exposition, 1904, 6
Starke Branch staff, photograph, 39
Starr, Rev. Frederick, 2

station organization
    cross-discipline, 68
    projects, 53
    research centers, 36
statistical research, 27–29, 41–42, 47
Stennis, John, 42–43
strip mine reclamation, 71
Sudworth, George B., 4–6
surface, environment, and manage-
    ment (SEAM), 71
Swank, Wayne, 81
Swanson, Fred, 81
Sylva: A Discourse on Trees, 2

tanoak studies, 6
Taylor, Raymond, 46
Taylor Grazing Act, 27
technology transfer
    Arnold, R. Keith, 65
    Buckman, Robert E., 72
    Dickerman, Murlyn B., 71
    GAO criticism, 64–65
    Graves, Henry, 9
    Roth, Filibert, 5
Terminologia Forestal, 60
termite control, 52
territories, forest research funded, 46
Theophrastis, influence on forest
    science, 2
Thomas, Jack Ward, 81–82
Three Sisters Wilderness Area, 58
3-Bug Program, 70
Tiemann, H.D., 6
Timber Conservation Board, 25
Timber Culture Act, 45
timber physics, 5
timber research
    Adams, John Quincy, 1
    Fernow, Bernhard Eduard, 5
    Forest Products Laboratory estab-
    lished, 8
    Harper, Verne L., 45, 47–48

Timber Research Review, 47–48
tool die models, 52
Train, Russell, 66
translations, CIA-funded, 60
Tratman, E.E. Russell, 4
tree identification, 4–5
tropical forest research, 81
tropical wood studies, 34
Trujilla Creek, photograph, 18
tussock moth, 57, 66, 70
2,4-D, 37
2,4,5-T, 37

UNESCO, 76
university-based forestry research,
   56–57
U.S. Agency for International
   Development (USAID), 35, 76–77
U.S. Geological Survey, 3–4
USAID (U.S. Agency for Inter-
   national Development), 35,
   76–77
USDA Competitive Research Grants
   Office, 74
USDA relations, 40–42
*Utilization of Hardwoods Growing on
   Southern Pine Sites,* 67
Utilization of the Southern Pines, 67

Varela, A.G., 11
*Variations in Naval Stores Associated
   with Specific Days Between Chippings,* 27
von Schrenk, Herman, 6

Wadsworth, Frank H., 76–77
Wagon Wheel Gap Experiment Sta-
   tion, 9–10, 84
Wakely, Philip, 16, 23, 31
War Production Board, 33–34
war-related research
   Cold War, 51–52

World War I, 12–14
World War II, 33–34
Watershed A and B, 51
Watershed and Aquatic Habitat Re-
   search, 65
watershed research
   Coweeta Experimental Forest, 51,
      81, 84
   forest/flood studies, 9–10
   Forsling, Clarence, 51
   Harper, Verne L., 50–51
   Hayden, Carl, 50–51
   Hursh, Charles R., 51
   Rocky Mountain Forest and Range
      Experiment Station, 50–51
   scope of, 83–84
   Swank, Wayne, 81
   U.S. Geological Survey, 3–4
   vegetation management, 67
   Wagon Wheel Gap Experiment
      Station, 9–10
Watts, Lyle, 32, 35
Weber, Barbara, 80
Weeks Act, 9–10
The Western Range, 26–27
White House science staff, 54
Whitten, Jamie Lloyd, 39
Whitten Bill, 40
Whitten report, 40
Wickard, Claude, 35
wilderness areas, 58–59. *See also*
   recreation research.
Wilderness Bill, 58
Wildlife Habitat Ecology Research,
   65
wildlife research
   Arnold, R. Keith, 65
   Harper, Verne L., 43
   Jemison, George, 60
   Thomas, Jack Ward, 81
Wind River Experimental Forest, 18
wind speed gauges, car-mounted, 29

Winslow, Carlisle P., 8
Winters, Robert K., 44
women, hiring, 58, 80
wood chemistry, 8
wood conservation, 4
wood decay
   Fernow, Bernhard Eduard, 5
   Forest Products Laboratory, 81
   Roth, Filibert, 5
Wood Handbook, 29
workforce diversity, 57–58, 80
World Bank, 35
World Congress, 76
World War I, 12–14
World War II, 33–34
W.R. Grace Commission, 74

Wray, Robert, 53
Wyman, Harper, 27
Wyman, Lenthall, 27

Zectran, 65–66
Zivnuska, John, 47
Zon, Raphael
   approving station locations, 7
   Central Investigative Committee, 8
   commercial tree study, 6
   forest/flood studies, 10
   Lake States Experiment Station, 15
   Office of Silvics, 8
   photograph, 62
   recreation research, 31